CONTENTS

Making the case for health and safety

Element 1

EXAMPLE

The human cost

In November 1990, Ken Woodward was involved in an accident at the Coca-Cola Schweppes UK factory in which he worked. The company had run out of a pre-mixed substance and so Ken had to mix up two chemicals himself. There was no safe system of work in place for this procedure, and no training had been provided by the employer, so Ken handled and mixed chemicals in a dangerous manner. The resulting explosion caused Ken to go blind.

An investigation revealed that there had been at least two previous near misses by workers carrying out the activity in this way. It was clear that these were not properly investigated and had been blamed on 'operator clumsiness'. A memo suggesting that the process should be changed for safety reasons was still sitting on the relevant manager's desk waiting for action when Ken had his accident. Thankfully, a colleague had helped Ken into a safety shower, which had only been repaired by a maintenance engineer the previous day. The safety shower had been out of action before then, but had not been reported as faulty and in need of repair. The maintenance engineer saw the damage and decided to repair it. If he had not, it is likely that Ken Woodward would have died.

As well as being fined, the accident cost the company nearly £3 million. After this accident, Coca-Cola Schweppes changed their approach to safety, and Ken became an advocate of safety standards and practices for the organisation. In 2006, Queen Elizabeth II awarded him the OBE medal in recognition of his services to health and safety.

Find out more at **www.kenwoodward.co.uk/kens-accident**

Example: The human cost

Have you ever had an accident at work or witnessed somebody else have an accident?

Think about the impact an injury, and time off work, would have on:

- you;
- your family;
- your colleagues;
- your work activities; and
- your organisation.

Feedback/Conclusions

An injury and time off work could mean your income being affected during your time off, which may lead to debt, especially if you are absent from work for a long time. It may affect all areas of your life, for example not being able to exercise, socialise with friends or play with your children.

Your oganisation may face legal action or compensation claims as a result of your injury or ill health.

Activity

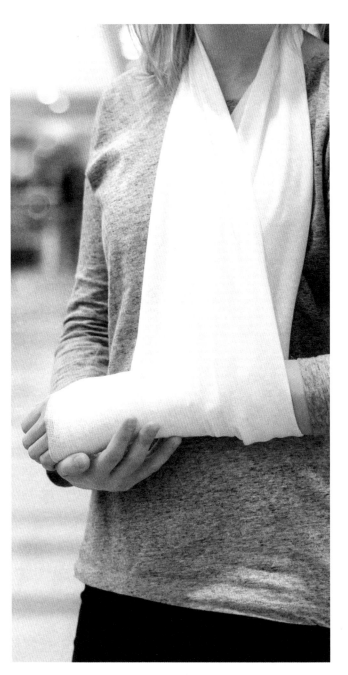

1.1 The moral, legal and financial reasons and benefits for managing health and safety

Most companies have limited resources, in terms of money, people and time. So sometimes it can be hard to convince management to spend money on health and safety. There are three compelling arguments that can be used to persuade them that managing health and safety is the best thing for business:

- moral;
- legal; and
- financial.

The moral reasons: 'doing the right thing'

Businesses don't usually injure their workers on purpose. They try to 'do the right thing'. This is known as a moral, ethical or humanitarian reason.

Society no longer accepts that people should be harmed or caused ill-health by their work activities.

The International Labour Organization (ILO) estimates that about 2.3 million people around the world die each year from work-related accidents and diseases. That's more than the average annual deaths from road accidents, war and HIV/AIDS combined.

This total is just for deaths. A much larger number of people suffer non-fatal injuries and illness at work. Plus, of course, injuries and illness don't just affect the workers; their families and friends are affected too.

Many – if not all – of these injuries and diseases are preventable.

The legal reasons: avoiding punishment and compensation

Moral principles often end up being written down as laws. Breaking laws can have serious consequences, and health and safety law is no exception.

Most countries have laws that punish businesses that don't do enough to protect workers (and others such as contractors and visitors) from getting injured or becoming ill at work. Punishment can take the form of fines or even prison. Businesses may also have to pay financial compensation to the injured person.

Organisations that comply with the law are less likely to face these claims.

The law places most of the weight of responsibility with the organisation, as their activity creates most of the risk. Organisations have a duty to take care of the health, safety and welfare of their workers (when at work) and also of other people – such as visitors or members of the public – who might be affected by their activities.

But the law also recognises that many other parties need to work together to ensure acceptable standards in the workplace, for example:

- individual workers (including supervisors and managers);
- landlords of workplaces; and
- designers, manufacturers and suppliers of equipment and substances used for work.

Enforcing the law

The law sets minimum standards, but it isn't very useful unless it is enforced. Health and safety law is enforced by regulatory authorities (sometimes called enforcement agencies or labour inspectorates). They employ enforcement officers (or labour inspectors) that visit organisations. One of their jobs is to check that organisations are complying with the law and take action when they are not. The legal action they can take can vary depending on the country, but here are a few common approaches:

- **Enforcement notices** – these are formal documents issued by the enforcement officer. These can be issued immediately, with no need to wait for a long legal process. There are two basic types of enforcement notice:
  - The first type requires an organisation to make improvements within an agreed period of time (eg, make improvements to segregation of vehicles and pedestrians in a busy warehouse).
  - The second type immediately stops a dangerous activity (eg, a dangerous piece of machinery must not be operated until it has been made safe).

If the organisation ignores these notices there can be serious consequences, for both the organisation and the individual (such as prosecution).

- **Criminal prosecution** – this is where the person or organisation is taken to court by an enforcement agency. This is the ultimate sanction imposed by the enforcing agency. It might be used if all other enforcement action has failed to improve conditions in the workplace, or can be used straight away if the issue is serious enough.

Successful criminal prosecutions will usually lead to some form of punishment. This might be a fine for an organisation, or a fine and/or imprisonment for an individual.

The criminal prosecution process takes a lot longer than issuing an enforcement notice.

Moral, legal and financial

The financial reasons: saving money

Health and safety costs money to put in place but it can also save a lot of money by avoiding costs from accidents. That means good health and safety is also good business.

We've already seen that not obeying the law can cost money (fines and compensation). But while some costs are obvious, others are 'hidden'. For some costs – like the cost of damage to reputation – it's difficult to estimate in purely financial terms.

Here are just a few of the many costs:

Financial costs
- Sick pay (paying the injured worker whilst they are off work).
- Lost time (production has to stop).
- Repairs to damaged equipment.
- Legal costs (paying lawyers, fines to the courts, compensation to injured people).
- Accident investigation costs.
- Increased insurance premiums.

Non-financial costs
- Reputation with customers.
- Reputation with neighbours.
- Poor staff morale.
- Bad publicity.
- People being less likely to want to work for an organisation.

Major accidents obviously have large costs, but even small incidents can have larger costs than you might think.

Costs can grow

A fire at the DuPont facility in Texas, USA, created the following costs:
- $724,000 fine to the company to settle clean air act violations;
- $6.8 million improvements made to chemical facility after regulator recommendations.

These are just the costs that we know and can calculate. There was also a great deal of disruption to businesses operating in the area. For example, major roads in the area had to be closed while emergency services dealt with the incident. Many of the costs associated with a major incident such as this are more difficult to quantify.

For more information, visit: **www.hse.gov.uk/pubns**

It is clear that many of the costs after something goes wrong are far larger than the cost of any safety measure to have prevented an incident in the first place.

There are very strong, convincing arguments for organisations to effectively manage health and safety. We will look at how this can be done in more specific areas in later elements of the course.

The Health and Safety Executive (HSE) is the Health and Safety Regulator in Great Britain; its website is a very useful source of free guidance not only on UK law, but also on best practice.

EXAMPLE

Imagine one of your work colleagues has been seriously injured after being hit by a forklift truck on a pedestrian walkway. Investigations have found that the forklift had not been maintained in accordance with the manufacturer's recommendations.

Consider the moral, legal and financial implications for both your colleague and the organisation.

Health and safety roles and responsibilities of relevant parties

Individual responsibility

A worker is working at height on a scaffold platform. They are wearing a hard hat and hi-vis, but there are no guard rails in place, or anything to stop them falling from the platform. Who may face prosecution in this example? Discuss with your group and report back to the class.

A scaffolder was given a 26-week jail sentence, suspended for 12 months, for working on an 18m-high platform wearing an unattached harness. If the scaffolder had fallen onto the concrete deck of the car park, it's highly likely that his injuries would have been fatal.

In this case, the scaffolder was personally prosecuted, not the organisation that he worked for. This is because health and safety law obliges workers to take reasonable care for their own health and safety when at work. The investigation found that his employer had taken reasonable steps to avoid working unsafely at height: the worker was well trained and experienced, and had the correct equipment available to him in order to work safely.

EXAMPLE

We often think of only organisations being responsible for health and safety. This example shows that each one of us is also individually responsible. Recklessness has consequences.

Roles and responsibilities of organisations

The prime responsibility for health and safety at work lies with the employer. This covers responsibility to workers, and extends to others who may be affected by the work activities.

Responsibilities to workers

Employers are responsible for the health, safety and welfare of their workers. This broad duty usually includes the provision of:

- safe plant/equipment;
- safe systems of working (ie, the way the activities, people, equipment, etc. are organised);
- adequate instruction, training, supervision and information;
- a safe place of work; and
- basic welfare facilities (eg, drinking water and sanitation).

How far an employer should go to perform these duties (ie, what is considered 'reasonable') can differ. It is generally accepted, however, that what is reasonable is a balance of the risk of injury/harm against the cost, in terms of time, effort and money. Most people would agree that it would be unreasonable to spend a huge amount of money to make something only a little bit safer.

The improvement in safety has to be sufficient to justify the cost, unless the law demands that actions are taken, whatever the cost.

For example, in a workshop, if painting is done once a year by two workers over a day, it wouldn't be considered 'reasonable' to install expensive, permanent ventilation equipment to protect them. The cost of this risk control would greatly outweigh the benefits.

Instead, a cheaper, temporary solution (but still one which gives reasonable protection) might be considered, such as:

- ensuring the area of work is well ventilated;
- that other workers not involved in the work activity are kept out of the area whilst the work is undertaken; and
- the worker is provided with a tight-fitting face mask that gives the correct level of protection from the paint being used.

Responsibilities to others

As well as duties to their own workers, employers have a responsibility to protect the health and safety of others who may be affected by the business activities, including:

- employment agency workers;
- contractors;
- visitors; and
- members of the public.

Workers may be very familiar with the risks in their own workplace. But a visitor or member of the public may not have any appreciation of the dangers associated with a particular place of work.

Warning signs are commonly used to warn visitors and members of the public of the hazards posed on sites. It is also usual to provide information to visitors when they arrive on site, for example site rules and emergency action.

Moral, legal and financial

Roles and responsibilities of directors, managers and supervisors

In most medium and large organisations, the 'employer' is the organisation itself. In practice, the responsibility for fulfilling the employer's duties for health and safety will fall on the management of the organisation. Health and safety isn't just the responsibility of the health and safety department, but is also a line-management responsibility.

Particular responsibilities for health and safety are placed on directors and senior managers.

Directors

Directors and senior managers are rarely involved in the day-to-day running of an organisation. They give an organisation its direction and set its priorities, and essentially set the tone for health and safety throughout the organisation; it should be a regular topic discussed at senior level. All directors – both collectively and individually – have ultimate responsibility for ensuring the proper conduct of the company. In many countries they can be prosecuted or even banned from being a director if they don't carry out their duties properly.

Managers

Managers are responsible for the health and safety of the workplace areas and workers under their control.

Supervisors

Supervisors are responsible for the day-to-day implementation of health and safety policies and procedures that fall under their area.

Responsibilities of workers

The law usually expects that when a worker is at work they must:
- take reasonable care of their own health and safety and that of others who may be affected by what they do (or fail to do) at work;
- co-operate with their employer to help them fulfil their legal duties; and
- notice that it's not just what you do, but also what you should have done. So, a worker not wearing protective clothing, or checking machinery that they are responsible for checking, might be liable to prosecution.

Responsibilities of the self-employed

The self-employed have a duty to protect:
- their own health and safety (because they employ themselves);
- that of anyone else they employ; and
- that of others who may be affected by their work.

So, in practice, the self-employed may well find themselves doing many of the things that an organisation would do (but on a smaller scale).

Responsibilities that apply to everyone

Individuals (whether that's a worker, member of the public, visitors, or a self-employed person) who deliberately interfere with, or misuse, anything provided in the interests of health, safety or welfare, can be held legally responsible for their actions. For example, setting off fire-fighting equipment 'for fun' or bypassing a machinery guard on a piece of equipment would fall into this category.

Moral, legal and financial

Think about the duties we have looked at for employers and workers. Write down examples of where, in your work activities, you have seen evidence of your employer's duties to you as a worker.

An employer needs to adequately control the risks from their work activities, and workers must work in a way that keeps themselves and others safe.

Activity

Study questions

What are the three main reasons for effectively managing health and safety?

What effect would poor health and safety have on the likelihood of accidents occurring?

What can happen to an employer if they ignore their health and safety responsibilities?

Name two financial costs arising from accidents and ill health.

Name two non-financial costs arising from accidents and ill health.

1.2 Managing health and safety consistently well

Good health and safety management has a number of key ingredients, and we are going to look at these in more detail throughout this section.

It's crucial for organisations to have effective health and safety management in place when planning their work activities. It's not enough to just train workers on keeping themselves safe; the work activity itself (how and where it is to be done and equipment to be used) should be carefully planned from the start, to make it as safe as possible.

Many management systems are based around the principle of Plan, Do, Check and Act, which helps an organisation achieve a balance between systems and behaviours; treating health and safety management as an integral part of good management, rather than as a stand-alone system.

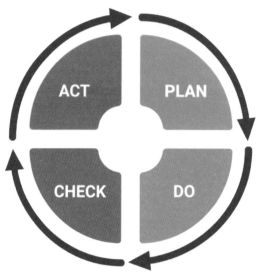

Plan	Determine policy/plan for implementation.
Do	Profile risks/organise for health and safety/implement the plan.
Check	Measure performance (monitor before events, investigate after events).
Act	Review performance/act on lessons learnt.

Organisations that are committed to effective health and safety management are less likely to have things go seriously wrong. They are also more likely to learn lessons from when things do go wrong, and improve safety to prevent a similar thing happening again.

Organisations can improve the way they manage health and safety by addressing the following areas:
- leadership – committed, promoting a culture that supports health and safety;
- policies and objectives;
- sufficient resources – in terms of equipment, materials, money and people;
- effective processes – including identification/control of risk and accident investigation;
- checking/monitoring performance – recognising opportunities and continually improving;
- communication;
- involving workers.

Some of these 'ingredients' are discussed in greater detail below.

Leadership

Management commitment starts at the very top of the organisation. Senior managers must inspire and motivate managers at all levels to treat health and safety objectives seriously and as equally important as other production-oriented objectives; this will help to promote positive health and safety culture in the organisation. You can do this by:

- Setting clear priorities and targets which follow from the organisation's safety policy, and making sure that they are met throughout the organisation.
- Leading by example – if senior managers are rarely seen on the 'shop floor', or don't take an active interest in health and safety, its importance is undermined. The best way for senior managers to show that health and safety is important is to lead by example by, for instance:
  - being approachable – encouraging workers to raise health and safety issues (and to treat those issues seriously);
  - encouraging ownership and participation in health and safety; and
  - treating company health and safety rules seriously – following company rules as well as enforcing them.
- Promoting changes to improve health and safety:
  - raising the status of health and safety committees, and health and safety professionals;
  - providing enough resources (equipment, people, time, money, etc.) to carry out jobs safely; and
  - providing appropriate training to improve skills and awareness.

Policies and objectives

It is important that organisations ensure they have appropriate systems in place to manage health and safety, including how to risk assess work activities, and how to investigate accidents and incidents.

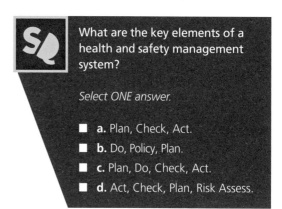

What are the key elements of a health and safety management system?

Select ONE answer.

- ■ **a.** Plan, Check, Act.
- ■ **b.** Do, Policy, Plan.
- ■ **c.** Plan, Do, Check, Act.
- ■ **d.** Act, Check, Plan, Risk Assess.

Managing health and safety consistently well

Effective processes

All organisations need effective systems to identify hazards, assess risks, and control those risks to an acceptable level.

Hazards can be identified in a number of different ways, for example checklists, task analysis, asking people who carry out the task, or referring to industry and official guidance. Having trained, capable people (whether managers, supervisors or workers) who can carry out the hazard identification process is essential.

Once identified, the risks from these hazards can be assessed. Risk assessment is a handy way to help you decide whether you need to do more to control risks in your workplace.

In most cases, the starting point for deciding if you have done enough is comparing the precautions you already have in place with those considered to be 'best practice' (usually written down in official guidance or an industry standard). It's very important that organisations learn lessons when things go wrong so that they can improve. So, another important process is incident investigation (which is covered in Element 3).

Checking/monitoring performance

Checking or 'monitoring' health and safety performance allows you to track and prove how well you're doing and spot opportunities to continually improve. There are two basic approaches to monitoring:

- Active monitoring – which is done to try to detect possible problems (and learning points) before something goes wrong. For example, workplace inspections to look for anything that may go wrong and identify actions to prevent possible incidents.
- Reactive monitoring – which is done after something has gone wrong. For example, where accidents and other safety-related incidents are investigated to find out what went wrong and identify actions to prevent a recurrence.

Active monitoring: inspections and audits

Inspections and audits are just two very common active monitoring methods. Both are effective ways of finding out what is actually happening (rather than what you think is happening). In some respects both are quite similar. But, in general, inspections tend to be a physical check of a workplace (such as an office or warehouse) or machine, often using a checklist of hazards/issues to look for. Audits tend to be checks of processes, procedures and systems, so will compare what should happen (according to a process/procedure) with what actually happens. Audits will look at evidence such as documents (records, written procedures and policies) and 'interviews' (asking people who are doing the job or those affected by it). Audits can be big (covering the entire management system). Most of the time they are more focused (covering a specific process such as how you learn lessons, how new risks are identified or how accidents are reported).

Reactive monitoring data

Data on accidents, dangerous occurrences, ill health and near misses has a significant part to play in monitoring safety performance. The benefit of this data is that it is easily understood by workers and can be compared with other organisations'/national data.

But reliance on accident data needs to be treated with caution. Here are just some of the many reasons why:

- The data only tells you what has already happened and does not necessarily tell you what may happen in the future. Very serious accidents tend to be quite rare, and it is difficult to predict when or if they will ever happen.
- Many ill-health problems take a long time to appear (maybe after years of repeated exposure to a hazard); when ill-health is first reported, it may already be too late to prevent other occurrences.
- Not all incidents get reported – there is often considerable underreporting of incidents – particularly of minor injuries and near misses.
- Injury statistics do not tell you how bad it could have been. For example, the data may record a cut finger from a dangerous machine, but it could so easily have been an amputation. It was just a matter of luck that it wasn't more serious.

Communication

Effective communication is essential for just about everything. It's also true of health and safety. For example:

- Individual workers must understand the hazards and risks of the work that they do, and all of the precautions they need to follow in order to keep themselves, and others, safe.
- Contractors working for an organisation must understand what the organisation expects of them – the rules that they are working under when they come on site. People are more likely to comply if they understand the reasons why they are being asked to do what is required of them.

All of these things require good communication. Communication can be in written form, or it might be verbal. Messages are sometimes communicated visually, either through the use of a picture or pictogram, or through a demonstrated behaviour. These communication channels may be formal or informal; however, the same health and safety message should be conveyed by all of the various channels and methods available to the organisation.

Some methods of communicating health and safety information include:

- safety signs;
- posters;
- tannoy announcements;
- safety briefings and toolbox talks;
- induction training sessions;
- health and safety committee meetings and minutes of meetings;
- formal company reports;
- emails and media postings;
- hazard alerts and bulletins;
- job appraisals;
- intranet.

Organisations must also take into account the difficulties that individuals may have in understanding the message that the organisation has tried to communicate. If the message is not received, or if it is misinterpreted, then communication has failed. Individuals may experience difficulties for various reasons. A hearing-impaired worker will experience difficulty if the message is conveyed verbally (eg, a tannoy announcement) or by sound (eg, an alarm). Similarly, a visually impaired worker will struggle to interpret text or a safety sign. A worker may be illiterate or a poor reader for medical reasons, such as dyslexia. Some workers may have a different first language, which could result in language barriers. The communication may have simply used terminology which a worker does not know the meaning of.

Whatever the barrier to good communication might be, the organisation must look for ways to eliminate or reduce the impact of that barrier.

Providing messages through varied media might be appropriate; for example:

- an audible alarm and a flashing light together;
- a picture or a pictogram on a safety sign rather than relying on text in one language;
- training documents in Braille;
- safety instructions available in multiple languages.

Consultation/participation of workers

There are some good reasons to involve workers in health and safety issues, especially those that directly affect them.

Workers doing the job will often know the hazards associated with their work activities better than anyone else and may have ideas on how to reduce risk. Consulting these workers can, therefore, lead to better solutions to everyday health and safety problems or even seeing opportunities that no one else could see. It is also more likely that they will follow policies and procedures that they themselves helped develop.

An employer does not have to consult with workers on everything (otherwise you'd never get any work done). Here are just some areas where consultation with workers would be needed:

- Introducing something (controls, processes, procedures, etc) that affects the health and safety of the workers concerned.
- Appointing people to specific health and safety responsibilities (eg to specifically help with health and safety duties or a role in emergency procedures).
- Any health and safety training or information the employer is required to provide to workers.
- The health and safety consequences of new technologies being planned or introduced in the workplace.

It is common practice to consult workers using a formal health and safety committee. The committee may be made up of worker representatives as well as management and health and safety professionals.

One benefit of a health and safety committee is that it acts as a focal point for health and safety issues. The health and safety committee also provides an opportunity for the involvement of workers in health and safety, which can improve motivation and the overall health and safety culture of the organisation.

 Why should you consider a range of communication methods when providing health and safety information?

Select all answers that apply.

- **a.** For variety.
- **b.** To take account of communication barriers.
- **c.** To ensure everyone is treated differently.
- **d.** Just in case someone misses a meeting.

Managing health and safety consistently well

Think about your organisation's health and safety policy statement. How is it communicated to you? How do you think it could be communicated better?

As you will see, every organisation has different responsibilities and priorities about health and safety, and they can highlight these specific targets in their policy statements. Lots of people are affected by an organisation's activities eg, visitors, contractors and workers. Unless you tell then what the hazards are and what is expected of them they won't know, so it's important to communicate properly with them.

Activity

Study questions

What are the typical responsibilities of employers to their workers?

What are the responsibilities of employers to people other than their workers?

What are typical responsibilities of workers?

What should employers consult workers about?

Stopping incidents and ill health before they happen

Element 2

The need to know something about the risk

A worker was injured when he was cleaning a machine in a factory. The factory manager had carried out the risk assessment for this task. But the manager didn't really know how the machine operated; he hadn't noticed that some of the safety guards and interlocks were missing or didn't work properly when he did the risk assessment. He hadn't been trained on how to complete the company risk assessment form either.

Think about how risk assessments are done in your organisation. Do you think the people who do the risk assessments know enough about the risks they are assessing? Do you think the right people are involved in the risk assessment process?

Risk assessment depends on people who know what to look for. They need to be familiar with the activity, equipment or situation being assessed. Sometimes this might involve a small team of people with different skills. But, if you don't know what to look for, you can easily miss significant hazards and, because you don't know they exist, do nothing to control them.

Activity

2.1 Risk assessment theory

What 'hazard', 'risk' and 'risk assessment' really mean

We risk assess every part of our day, often without thinking about it. For example, driving quickly across a busy junction. In this element, we look at a systematic approach to risk assessment. But first we need to look at some important terminology, so that we all know what we mean.

Hazards always exist to some degree in the workplace. The risks arising from a hazard can be controlled so that the likelihood of harm is reduced to an acceptable level, or sometimes eliminated completely. A risk assessment is the first key step in achieving this.

What work activities may require a specific risk assessment?

Select ONE answer.

- ■ **a.** The likelihood of something causing harm.
- ■ **b.** Something that has the potential to cause harm or damage.
- ■ **c.** Anything in the work environment.
- ■ **d.** The severity of an injury.

Key terms

Hazard
Anything that has the potential to cause harm or damage – this could be an object, an activity, or even a situation or a combination of these.

Hazards are often described either by referring to the type of harm or effect they lead to (eg, mechanical machinery hazards such as crushing, entanglement or even slips, trips or falls) or instead by the hazard origin or source (eg, electrical or noise).

Risk
The likelihood that a hazard will cause harm combined with how severe the harm could be.

Risk assessment
The process of recognising hazards, deciding if the risks are acceptable, and seeing what can be done to make things safer.

Control measure
Anything that you do or put in place that eliminates or reduces the risk of harm or damage.

Why we do risk assessments

The aim of a risk assessment is to make sure that no one suffers harm as a result of workplace activities. To do that, we need to look for things that might harm you and decide what to do about them. A risk assessment is a systematic way of doing that, so that we don't miss anything important. It helps you make decisions about whether you are doing enough in your workplace to control risks or whether you need to do more. It's important to use it correctly – it should help you make decisions – and it shouldn't be used in reverse (to justify a decision that you have already made!).

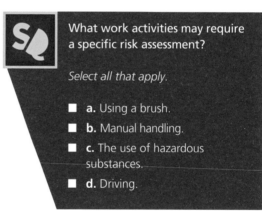

What work activities may require a specific risk assessment?

Select all that apply.

- ☐ **a.** Using a brush.
- ☐ **b.** Manual handling.
- ☐ **c.** The use of hazardous substances.
- ☐ **d.** Driving.

Risk assessment theory

How you do risk assessment

The process of risk assessment is not difficult. It's about us figuring out what might potentially go wrong, and what we need to put in place to stop or minimise the chance of this happening.

Although there are no fixed rules about how a risk assessment is conducted, it is important to take a structured approach so that all relevant risks from hazards are addressed. The leaflet 'Risk assessment – A brief guide to controlling risks in the workplace', issued by the British Health and Safety Executive (HSE), sets out a good framework for the process. The guide, called 'INDG163', can be downloaded from the HSE website.

This guidance uses a five-step approach:

The five steps to risk assessment

STEP 1 — **Identify the hazards**

STEP 2 — **Identify who may be harmed and how**

STEP 3 — **Evaluate the risks and decide on precautions**

STEP 4 — **Record your significant findings**

STEP 5 — **Review and update when necessary**

We look at each step in turn below.

Risk assessment theory

1. How to identify hazards

It's important to look at all hazards in the workplace to identify the most important ones.

Look at all aspects of work, including:
- the way in which the work is carried out;
- the way it is organised;
- the substances and/or equipment used; and
- think about what harm can be caused.

Some hazards are obvious – if you are climbing up and down ladders, you can slip or fall and injure yourself.

Occasionally the hazards are less obvious, particularly where a normally safe activity or situation is only hazardous sometimes. For example, faulty equipment (electrical equipment with damaged wiring may cause shocks or burns) or a specific scenario (a spillage would make walking across a floor hazardous).

So, you need to think not only about routine/everyday activities, but also occasional or non-routine activities such as maintenance work, loading and unloading of deliveries from vehicles, or changes in production cycles. Interruptions to the work activity can also be a common cause of accidents.

There are many ways of identifying such hazards, such as:
- Instructions from manufacturers and suppliers (eg written in the equipment user manual or chemical safety data sheet).
- Official guidance/law – not only from the health and safety regulator but also industry professional bodies. They will point out typical hazards in different types of businesses and activities.
- Records of past incidents (accidents, ill health, near miss) in your workplace. This will tell you how people can be harmed.
- Talking to people who actually do the job (as well as relying on your own expertise).

To help you look at hazards systematically, people often use hazard checklists (compiled from the above sources). These can prompt you as you inspect the workplace or analyse each step of a job or task.

2. How to identify people at risk and how they may be harmed

Identify who may be harmed and how it is important to not just think of those carrying out particular activities, but also of all those who may be affected by those activities. Consider how the following groups of people can be harmed.

Workers
Workers are those directly involved with the activity, or others working nearby in the workplace. These could include vulnerable people, which is discussed later.

Contractors
Contractors might not be fully aware of all the hazards or procedures at the place in which they are working. A risk assessment should include contractors and consider how to provide them with the same level of protection as your own regular workers.

Visitors and members of the public
Visitors are even more at risk than contractors; they will usually have no awareness of any of the hazards, because they aren't familiar with your workplace or what you do there. Don't just think about wanted visitors; think about dangers to 'unwanted' ones as well – children are often attracted to 'fun' places to play like building sites and railway lines.

Vulnerable people
This group includes anyone who is particularly at risk, such as the young, pregnant/nursing mothers or inexperienced workers, lone workers and staff with disabilities. Vulnerable people may be subject to different levels of risk, depending on skill level, experience, age, mobility, etc.

3. How to evaluate the risk and decide on precautions

After identifying the hazards and how people may be harmed, you need to work out if the risk of harm is significant or not and, if it is, whether you need to do more.

There is usually a limit to how much a risk can be controlled, so we need a way of deciding what the priorities for action are, as well as what is actually reasonable and practical to do. Our goal is to determine the appropriate level of protective measures to put in place for each hazard.

If you think about it, it is obvious that the same hazard can lead to different results (ie, different severities). Each result can have quite a different chance of happening (ie different likelihood).

For example, a carpet with holes in is a trip hazard. The risk of a serious injury through tripping and falling depends on things like where the hazard is. If the hole is behind a desk where there is no access, the risk will be low as people are unlikely to go there. If the carpet is on a flight of stairs which is regularly used but not well-lit, the risk of a serious injury is much greater. Some people use a risk rating matrix to assign likelihood and severity ratings to identified risks. But this is not necessary in most cases and, if it is used, it needs experienced people to use it – otherwise it is possible to get it quite wrong, ignore serious risks and focus on trivial risks. It is better to start simple and then only use more complicated methods if you need to.

For most common work situations, there is a much quicker way to decide whether you are doing enough. Check whether there are any national or international laws applicable in your region that require action on specific hazards (such as chemicals, machinery, etc.). This is helpful when conducting risk assessments because the hazards, and often the level of risk and appropriate control measures, have already been identified. There is usually official guidance available from regulators, industry associations or international standards. There might also be 'best practice' guidelines that your organisation has developed itself. Comparing what you have with what the guides/standards say will tell you whether what you are doing is enough and whether more is expected.

A word of caution though – just make sure that the guidance you are using actually applies to the type of work that you are doing (and the way you are doing it). If you are doing something unusual, those guides probably won't work for you and you'll have to think about it in much more detail and even consult specialists. We will look in more detail about approaches to choosing controls later.

Prioritisation of actions based on risk
When you've compared what you are doing with what is needed (eg, from guidance), you will usually have a list of extra things to do – these actions need to be planned (what, by when and by whom), prioritised and actually completed. It is recognised that a working environment completely free of any risk is a practical and financial impossibility. When choosing measures to put in place, we are trying to achieve a balance between the improved safety that they will bring, against the resources (time, effort and cost) involved in establishing these measures.

Resources are applied that have the most beneficial effect of controlling the risk. If you have been able to use official guidance, these issues will already have been taken into account in the guidance itself.

Priorities for actions obviously need to be based on the level of risk remaining after you have taken account of what you already have in place; the higher the risk, the higher the priority.

A timescale for action can then be applied. An example is shown in the following table (but you may well have a different system in your workplace):

Priority 1 (immediate)	Work should not continue until the risk has been reduced.
Priority 2 (short-term)	Measures should be taken within the next 14–18 days.
Priority 3 (medium-term	Measures should be taken within the next 6 months.
Priority 4 (long-term)	Non-urgent action is required within the next 12 months.

4. How to record significant findings

There is no generally agreed standard format or layout for recording a risk assessment. Whatever documents are used, ensure that they include the following:

- A description of the processes/activities assessed and identification of the significant hazards involved.
- Identification of any group of people at particular risk.
- Evaluation of the risks:
  - consideration of existing control measures and whether they are sufficient;
  - if you need to do more, assign responsibility for putting in place any additional controls.
- Date of the assessment and date for review.
- Name of the person carrying out the assessment.

5. The reasons for reviewing risk assessments

Risk assessments are based on what you knew when you did them. But things change. So risk assessments should be reviewed, especially if you believe there have been changes to information or circumstances which might affect the outcome of the assessment. For example:

- a change in legislation;
- a change in control measures;
- any significant change in work practices and processes;
- installation of new machinery and equipment.

Even if a re-assessment is not triggered by a change, it's good practice to review risk assessments periodically to ensure that they remain accurate and relevant. Many organisations review them annually as a matter of policy.

Don't think of a risk assessment as a one-off event. It is an ongoing process where work methods and safety precautions are monitored and kept under regular review. This ensures that:

- there is continued compliance with legislation;
- changes in equipment are taken into account;
- changes in materials are taken into account;
- changes to work methods are incorporated;
- systems still work safely in practice;
- opportunities to improve control measures are recognised;
- safety precautions are adjusted to take into account accident experience.

For an example of a good risk assessment form, take a look at the British HSE's risk assessment template on its website.

Controlling risks

When thinking about control measures, you'll usually have to apply a range of them and you'll often hear the term 'Hierarchy of Control'. This means that there is a preferred order of applying control measures because some control measures are more effective than others. But, as mentioned earlier, this sort of thing is already taken into account if you are using official guidance about what control measures need to be put in place. If it's more complicated than that, and you can't simply rely on guidance that applies to the work that you do (and the way you do it), you need to think a bit harder about what control measures you may need. And that's where the hierarchy comes in.

Ideally a risk will be eliminated, but there are other ways to control it if that's not an option. Using personal protective equipment (PPE) is at the bottom of the hierarchy, as it only protects the person rather than addressing the hazard.

Hierarchy of Control

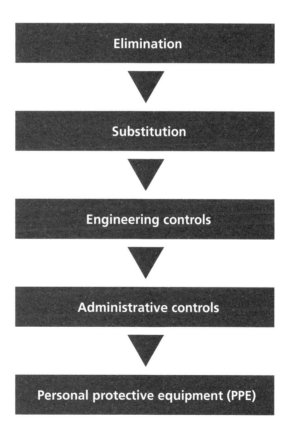

Elimination

Substitution

Engineering controls

Administrative controls

Personal protective equipment (PPE)

Risk assessment theory

The following is an outline of the Hierarchy of Control, in the order in which each measure should be applied.

Elimination

Wherever possible, remove the hazard completely:
- If a job involves significant risk of a manual handling injury, introducing a mechanical alternative will eliminate the risk (but may introduce other, different risks).
- If repairing or servicing an electrically powered machine, unplugging it or disconnecting it from all sources of energy (including stored charge) will eliminate the risk of electric shock.

Substitution

Replace the hazard with something that is less dangerous:
- By replacing solvent-based paint with water-based paint, the chance of fire and explosion is reduced.
- By substituting chemicals in powder form with pellets or paste, the risk of inhaling airborne dust is reduced.
- When working at height, replacing a ladder with a tower scaffold will reduce the risk of falling.

Engineering controls

Engineering controls involve designing safety into equipment and machinery, vehicles, containers, etc.

This could include putting guards on machinery and enclosure features, as well as adding special-purpose safety equipment to processes to remove or reduce risks.

For example:
- Totally enclosing a process that generates dust or fumes to prevent the escape of airborne contaminants that could be inhaled by workers nearby.
- A cover over a light switch.
- Installing perimeter fencing or other types of barriers around machines.
- Designing operating controls that are easy to see and use, not allowing machines to be turned on accidentally, and incorporating emergency stops.
- Reducing risk from of dusts and vapours by installing ventilation.

Administrative controls

Signage and warnings

The use of signage and warnings is important in the workplace. Safety signs can highlight hazards (eg, a sign warning of a wet floor) or give instruction ('No smoking' or 'Hearing protection must be worn'). Signs are also used to indicate safe places and conditions ('Fire exit').

Safe systems of work

> ### Key term
>
> **Safe systems of work**
> A formal procedure for a task that defines safe methods of working, to ensure that hazards are eliminated or risks minimised.

Safe systems of work are necessary when an element of risk remains in a work activity, so are often produced as a result of a risk assessment. They communicate the hazards, controls and the way in which the task should be carried out, along with any emergency arrangements needed.

Permit-to-work

> ### Key term
>
> **Permit-to-work**
> A formal written document of authority to undertake a specific procedure. A permit-to-work is intended to protect those working in high-hazard areas or activities.

A permit-to-work system is designed to ensure that all necessary actions are taken before, during and after particularly hazardous operations. For example, maintenance work can only begin when normal controls are temporarily removed (such as guards).

For other work, the hazardous operation is the main part of the job. In these circumstances, special precautions must be taken. Examples of this work include:
- Working with or maintaining high-voltage electrical equipment.
- Working with hot or highly flammable materials.
- Pipework containing hazardous substances.
- Working in confined spaces.

Work instructions

All workers (including contractors) must read, understand and correctly apply instructions.

All instructions need to be properly documented to provide a precise reference for all workers. For example:
- Short notes instructing what to do if the toner needs changing in the photocopier, displayed on a wall nearby.
- Manuals detailing what steps to take when carrying out more complex or lengthy procedures, such as calibrating and setting up grinding wheels for operation.

These form the basis of training programmes. They are likely to be accompanied by checklists for workers. They are used as aids to ensure that all the correct steps are taken, and to check off details before continuing with the next operation or starting operations up again.

Information, instruction, training and supervision

An important part of any control strategy is to make workers aware of the hazards associated with the work they are doing. They also need to know what control measures are in place, and what they are required to do, in order to keep themselves and others safe.

This can be achieved by:
- adequate instruction and training to create competent workers; and
- supervision (particularly important when the worker is inexperienced).

Providing information, instruction, training and supervision to ensure the health and safety of workers is usually a requirement of health and safety legislation.

Personal protective equipment

Key term

Personal protective equipment (PPE)
Equipment, clothing or accessories worn by a person at work, which protect them against risks to their health or safety, such as wearing protective gloves when using cleaning chemicals.

PPE protects a worker from one or more hazards. While useful as a control measure, it is limited in its effectiveness because it:
- doesn't address the hazard directly;
- only protects one person;
- only protects the person if it is used correctly; and
- can be uncomfortable or difficult to wear while doing physical work.

Because of these points, PPE is often not worn correctly, or worn at all.

PPE is therefore at the bottom of the control hierarchy, and is only used when other controls cannot be implemented or are not enough to reduce the risk down to an acceptable level.

Being sensible and proportionate

Risk assessments are not supposed to be the end result. They are a means to an end; a tool for helping you control risk. To get the best out of them and not get buried in paperwork, you need to be sensible and proportionate. Risk assessments are not required to be perfect but are required to be 'suitable and sufficient'. In real terms, this means:

- The risk assessment should identify all the significant risks in your workplace (concentrate on the important risks and ignore trivial ones – but it must relate to what is actually going on in your workplace – reality not fantasy).
- You're not expected to consider 'unforeseeable' risks – some are too remote or can't easily be foreseen when you do the risk assessment.
- The amount of detail needed in a risk assessment depends on the level and complexity of risk (if a risk is low, simple and easy to see, you don't need much detail at all).
- Once assessed, insignificant risks may not need further assessment or controls.

Your risk assessment can help you identify what risks and control measures to look at in more detail. Note that these control measures do not have to be assessed separately but can be considered as part of your overall risk assessment.

General versus specific risk assessment

A general risk assessment is where all of the various hazards presented by the work are considered, such as a workplace or work activity.

A more detailed risk assessment is sometimes needed where a significant or complex risk exists from specific hazards. Common examples are:

- hazardous substances;
- manual handling;
- laptops/computer workstations;
- fire.

Again there is no agreed format for recording these, so you can still record the significant findings in your general workplace risk assessment record.

Risk assessment theory

2.2 Risk assessment in practice

We are now going to put this theory into practice by doing a short risk assessment for a given work environment.

Identify the hazards in the augmented reality animation:

Complete a risk assessment using the form on the opposite page, carefully considering who may be affected by the identified hazards and what should be put in place to adequately control them.

Activity

Risk assessment template

What are the hazards?	Who might be harmed and how?	What is already being done?	What else do you need to do to control this risk?	Action by whom?	Action by when?

Review date:

When reviewing the risk assessment, make sure that you consider all hazards that may present themselves during the work activity. This means not just to the worker but to others who may also be affected by somebody carrying out the work. So, although we cannot see any workers in the above picture, consider who is likely to be working in this area and who else could be affected.

It is also important that you assign an appropriate risk level to each of the identified hazards, and that you clearly identify control measures within your assessment.

Activity

Stopping incidents from repeating themselves

Element 3

EXAMPLE

Avoiding future incidents

One hundred and ninety-three ferry passengers lost their lives in 1987 when the *Herald of Free Enterprise* sank on a routine journey from Zeebrugge, Belgium, to Dover.

There were failures with processes, overworked employees and an overloaded ferry among many other things. The incident was entirely avoidable and the organisation faced a massive backlash for the choices that they had made in the run up to the incident.

Example: Avoiding future incidents

Have you had any incidents at work which were minor, but could have been a lot worse, where there was a lucky escape? What was done to stop this from happening again and from being much worse?

If we think about the incident on the previous page, although there were many failures at many levels, what is clear is that the organisation was not prepared for an incident like it.

3.1 Why investigating incidents makes sense

Key terms

Incident
An undesired event that has caused or could have caused damage, death, injury or ill health. There are two main types of incident – accident and near miss.

Accident
An incident which resulted in damage, death, injury or ill health.

Near miss
An incident that could have resulted in damage, death, injury or ill health but did not, in fact, do so.

Incidents happen every day, across all industries. It is important that we know what to do in the event of an incident at work, so that we can take immediate action, learn lessons, improve safety and stop the same thing happening to anyone else.

In Element 1, you learned how accidents may cause injury and time off work. This often results in a reduction or loss of income. The effects to people's short- and long-term quality of life must be considered.

Near misses occur even more regularly than accidents. If unreported, they can become accidents (because the causes are not dealt with), so it is vital that we learn from these situations too.

Far too many investigations just look at the obvious, immediate causes. You need to go deeper and find the root causes.

3.2 A simple four-step approach to investigations

There are four basic steps in the investigation process of an incident:
1. **Gather the information.**
2. **Analyse the information.**
3. **Identify risk control measures.**
4. **Develop and implement the action plan.**

These steps might be done together or repeated many times. For example, it is normal to gather some evidence which might give you your first ideas of how the incident happened. You then might need to gather further evidence to explore those ideas (to confirm or discount them).

1. Gather the information

Relevant information from the incident can be gathered in several ways:
- Physical evidence (at the scene, CCTV footage, etc).
- Witness testimonies.
- Documentary evidence, such as risk assessments, training records, or organisation policies and procedures.

It is important to capture information as soon as you can. If necessary, stop the work and keep everyone out of the area. This will reduce the chance of evidence being tampered with, like equipment being moved, guards being replaced, etc.

Talk to everyone who was close by when the event happened, especially those who saw it or knew anything about the conditions that led to it. The level of the investigation (in terms of scale and amount of time spent gathering information) should be proportionate to the seriousness (or potential seriousness) of the incident.

Collect all available relevant information. That includes opinions, experiences, observations, sketches, measurements, photographs, records and details of the environmental conditions at the time, etc. This information can be recorded initially in note form, with a formal report being completed once this stage of the investigation is concluded. Keep your notes at least until the investigation and report is complete, so that they can be referred to if required.

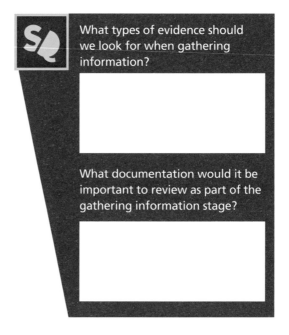

What types of evidence should we look for when gathering information?

What documentation would it be important to review as part of the gathering information stage?

2. Analyse the information

Key terms

Immediate cause
The obvious cause(s). For example, the exposed blade on a machine that has leads directly to a cut hand.

Root cause
The ultimate cause(s) that allowed the immediate and other causes to exist – these are mainly organisational and management failings; for example, the workshop was understaffed for the targets they had been set.

An analysis involves examining all of the facts. Look at what is relevant to the incident, in a timely, structured way. Consider:

- Is there any information missing?
- If so, why?

The structure of an investigation is helped by establishing a timeline of events, and looking at any unsafe conditions (eg a slippery floor) or unsafe acts (eg removing a guard) in the lead-up to the incident.

All possible causes and consequences of an incident should be considered during this stage of the investigation. There are usually multiple causes for an incident, which all come together. They all form a number of 'causal chains' where something causes a consequence and that is then a cause of something else. The chain stretches from immediate causes all the way back to the root causes.

Rooting out causes using the 5 Whys

The '5 Whys' strategy is a very effective problem-solving tool. It is used as part of an investigation to find out the reasons behind a problem. By asking a sequence of 'Why?' questions (five is a good rule of thumb as it is usually enough to take you to the root cause), we can work through the layers of issues that may surround a problem. This method can be used to identify the causes of an incident.

What is the main purpose of the 5 Whys strategy when investigating an incident?

Select ONE answer.

- ■ **a.** Reduces risk.
- ■ **b.** Identifies causes of an incident.
- ■ **c.** Produces action plans.
- ■ **d.** Identifies control measures.

A simple four-step approach to investigations

Example based on HSE's HSG245 workbook
A worker (David) discovered a problem in the edge machine. This is a machine which saws off and planes wood. He didn't know who to report the problem to so he opened the machine lid (which was connected to an interlock which automatically turned off the power when the lid is opened). David put his pencil into the interlock switch, so he could still operate the machine with the lid open. The saw suddenly started and cut David's hand.

Use the 5 Whys strategy to work through the incident and identify the issues that may have played a part in causing the incident. We have started you off below:

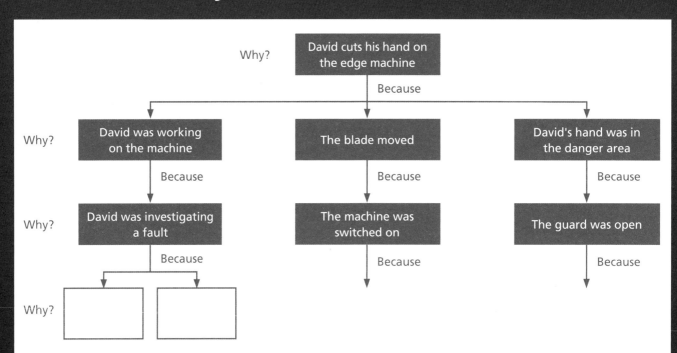

3. Identify risk control measures

The next step in the process is to review the risk control measures that are in place. This will determine whether the risks are adequately controlled within the work activity or if more needs to be done to keep people safe. Sometimes you may find that the risk controls are good enough but on this occasion had not been used. You may find, however, that more needs to be put in place to achieve the required levels of safety.

Your review should include the following:
- Considering if there are legal implications of the risk controls in place, and advise on changes needed.
- Checking if existing risk controls are sufficient, but were just not used or followed in this instance (whether someone decided not to or was unable to).
- Evaluating risk controls on their ability to prevent recurrence of a similar or worse event.
- Prioritising what needs to be done first – this might be to address something that is still dangerous, or actions for a routine task where the incident could easily recur.
- Creating an action plan with realistic timescales.

The most effective risk controls are those that create a safe or safer place of work, rather than rely on 'safe' people (eg wearing PPE). This is known as the Hierarchy of Control and is illustrated in Element 2.

We have already used this Hierarchy of Control in relation to risk assessment in Element 2, and the exact same approach should be used when reviewing existing risk control measures after an incident has occurred. Applying the hierarchy will help you to create as safe a situation as it is possible to achieve, by considering measures that:
- eliminate the risk before all else (inherently safe products);
- adequately control the risk at source (e.g. guards, local exhaust ventilation); or
- minimise risk by reliance on human actions, such as PPE.

You should also think about whether similar risks might exist elsewhere in the organisation. This can help to apply lessons learned from an incident:
- Within different departments of the organisation – information should be shared between staff, particularly supervisors or those responsible for safety.
- In different locations facing similar hazards – on different sites, control measures should be implemented throughout the organisation.
- In different organisations – how do others manage this risk? Sharing best practice through industry forums can help organisations learn lessons from each other.

If an organisation has had similar incidents prior to this adverse event, they will need to determine whether they were investigated thoroughly and why it has been allowed to happen again.

Organisations must not ignore safety failings. This would be included during a prosecution or investigation and could affect the severity of any penalties issued.

4. Develop and implement the action plan

The action plan should contain information about the controls used to deal with all causes identified.

Thinking about the same example (the worker who cut his hand), identify actions to prevent this from happening again and decide on a realistic timescale for getting these actions done.

Reflection

After an incident, organisations will need to ask themselves:
- What did the incident cost? Remember, costs can be money, time, resource or reputation.
- Who needs to know about these findings and actions? How will they be communicated?
- How are the actions going to be tracked and closed out?
- Are there any trends that require further investigation?

Organisations should keep records of adverse events, their causes and the actions taken to address them, to prevent anything similar happening again. These records will help identify any trends over time in order to improve an organisation's overall understanding and management of risk for its work activities. Including the cost of incidents will also help an organisation to appreciate the value of preventing accidents and ill health.

An incident will have many causes. During an investigation, we may also discover things that have not directly caused the incident, but need to be addressed to improve the wider health and safety within the workplace. It is important that we learn the relevant lessons from investigations and work to prevent the same thing from happening again.

Having effective processes in place for how and what to investigate will lead to a more thorough and effective investigation. Ensuring that all steps of an investigation are covered will help you fully understand how and why something went wrong.

Dealing with common workplace hazards

Element 4

EXAMPLE

Hazards working at the quayside

A worker was operating a forklift truck in the docks. His job was to move containers from a holding area to the quayside so that they could be loaded onto a ship. His view was restricted by the forklift mechanism. Another worker (a tally clerk), who was wearing high visibility clothing, was checking the containers at the quayside. As the forklift truck driver turned to check his clearance, the tally clerk stepped in front of the mechanism and was struck on the head by it. Although he suffered serious head injuries, luckily he recovered fully.

The incident investigation found that:
- work at the quayside was badly organised;
- arrangements to separate vehicles and pedestrians were inadequate;
- there wasn't enough space to manoeuvre the forklift truck;
- there was no co-operation between the forklift truck driver and dockside contractors;
- the work wasn't effectively supervised.

Workers need to understand hazards in a workplace and know what to do to keep themselves safe.

Example: Hazards working at the quayside

4.1 General workplace issues

Workplace access and housekeeping

You don't just need a safe workplace, but also safe access to it and, especially in an emergency, a safe way to get out. This may not seem much of an issue if you work in a building like an office, where access is permanent and 'built in'. But if you are working in a constantly changing environment like on a scaffold or in the cab of a tower crane on a building site, you need a safe way to get there too. Otherwise the most dangerous part of your day can be getting to work.

All walkways and corridors between working areas and leading to the outside of buildings should be free of obstructions.

Many accidents are caused by people colliding with other people, equipment or vehicles when coming out of a door. You can prevent these by having:

- one-way systems through double doors;
- automatic doors that can be easily opened when carrying or pushing loads; and
- hazard signals and warning lights on moving vehicles.

Housekeeping isn't just about keeping things in order, clean, tidy and well-maintained. It's sometimes described as 'a place for everything and everything in its place'. If you neglect housekeeping, sooner or later you will get more slips, trips and falls (because of clutter, leaks and spillages) and a fire could start and spread more easily by, for example, incompatible materials being stored too close together. All workplace areas need to be regularly cleaned, waste removed, spills reported and cleaned up, and pedestrian and traffic routes outside kept clear.

Lighting

Poor lighting can lead directly to incidents. For example, not being able to see a box in a corridor, because of low light, means you are much more likely to trip over it or bump into it. If being able to distinguish colours is important (such as electrical work), low light levels (and some types of light) can make this very difficult. Some types of light (flickering or strobe lights) can make moving machinery look like it has stopped – so you think it's safe when it isn't. You need enough lighting and it needs to be the right type. Natural daylight through windows is best but usually artificial lighting is also needed.

Different tasks also need different levels of light. For example, just moving around the workplace between rooms or corridors probably doesn't need much light at all. But fine, detailed work or work with a high degree of danger often needs more intense lighting. Since mains electricity can fail during an emergency, you also need independent emergency escape lighting too so that everyone can still find their way out.

Light shining directly into your eyes can cause discomfort, so lighting must be designed to minimise glare (eg fitted with a shade). Reflections can also be a problem, so angle computer screens to avoid reflections and avoid putting workstations facing windows.

Temperature

Workplaces need to be at a comfortable temperature. If they aren't, you can be distracted, lose concentration and be more accident prone. In extreme cases, it can be life-threatening. But what is comfortable (and how you achieve it) depends on what you are doing. There are general guidelines that suggest workplaces should be at least 16°C for people doing low activity work, such as working at a desk in an office, and at least 13°C for people doing hard manual labour. There might also be a recommended maximum temperature over which people should not be working.

Some work needs to be in extreme temperatures, like working with molten metals in a foundry or working in cold stores. These can lead to conditions such as heat stress (if very hot and humid) or frostbite (if very cold). You can make these working conditions more comfortable by providing:

- cooling ventilation (such as air conditioning) for hot environments;
- insulation for floors where workers are standing on cold floors;
- the opportunity for workers to take turns to reduce the time each person spends in the extreme temperature;
- warm clothing and gloves for cold environments and heat-resistant clothing for hot environments;
- rest facilities (away from the extreme conditions) and hot or cold drinks.

Slips, trips and falls (on the same level)

When you think of falling you probably think of falls from height. But people can often slip, trip or fall on level surfaces. Typical situations include:

- wet or greasy floors caused by spills, cleaning, snow, ice, etc;
- uneven or loose surfaces caused by broken or poorly laid paving, edges of mats, holes in carpets, etc;
- obstacles on the ground such as trailing cables, small boxes, bags and cases, pallets left too close to walkways, etc.

These situations may be made more dangerous where visibility is restricted, like when coming out of a doorway or around a corner or if lighting is poor.

People are quite often distracted when walking around in the workplace; they may be unaware of possible hazards and may not notice things such as obstructions or steps. As we discussed earlier, good housekeeping is one way of reducing slips, trips and falls. Some other common ways of tackling slips, trips and falls include:

- stopping floors becoming slippery or dirty (by putting mats at the entrances, fixing leaks, maintaining machinery (to prevent leaks), organising tasks to minimise spillage);
- keeping floors clean and using the right cleaning methods (clean up spillages quickly, use the right type of cleaning regime for the flooring that you have);
- keeping floors in good condition (checking for loose or broken sections and getting them fixed);
- considering slip-resistant floors (for areas that are likely to get wet or slippery);
- having good lighting on walkways (especially when there are slopes or steps) and keeping them free of obstructions;
- considering slip-resistant footwear (if floors are normally wet);
- installing good drainage (so normally wet areas dry off quicker); and
- using warning signs (where the hazard can't be eliminated, to help people alert people of the danger).

General workplace issues

Welfare

Welfare is about the wellbeing of workers. When dealing with hazards, it's easy to forget the basics such as water, food, rest breaks, washing and toilets. Without considering these, workers can become tired, dehydrated and ill. You should provide workers with:

- drinking water;
- toilets, handwashing facilities and if necessary, showers – the more workers you have on site, the more of these you need;
- changing rooms and lockers for workers who have to change into special work clothing (such as overalls, uniforms, etc.); and
- places to rest and eat (with suitable seating, away from any unsafe areas).

First aid

At some stage, regardless of precautions, people do get injured at work. Mostly, these are minor injuries, but even so, you need to be prepared so that things are dealt with quickly and don't get worse. All workplaces should have a first-aid kit. It should be easy to locate, clearly identified and in a well-lit area. First-aid kits must be properly stocked, clean and maintained.

Larger organisations might have a first-aid room depending on the number of staff and the risks that they face in the workplace. This should be made available not only to workers, but also to contractors, visitors and members of the public. It should be large enough for stretchers to be used and moved around.

Organisations will usually have at least one trained first aider, who can administer first aid in the workplace, such as controlling bleeding, or even giving artificial respiration, until professional medical help arrives.

What first aid you actually need in a workplace depends on several things, such as:

- the nature of the work (what hazards are present? What types of accidents have happened before? It pays to be extra prepared for the types of injury you are most likely to encounter);
- the location of the workplace (if is it far from hospitals or difficult to get to, help could take a long time to arrive);
- shift work and worker distribution (make sure that first-aid help is available whenever and wherever people are working, including those who routinely travel for work);
- cover for absence (illness or holidays); and
- visitors who might need first aid (including members of the public, who might be customers of your organisation).

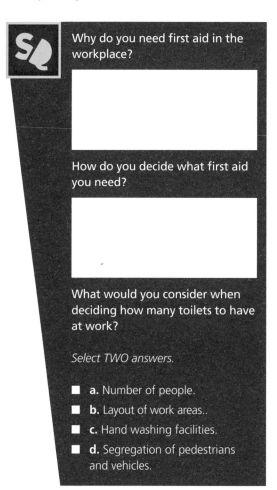

Why do you need first aid in the workplace?

How do you decide what first aid you need?

What would you consider when deciding how many toilets to have at work?

Select TWO answers.

- a. Number of people.
- b. Layout of work areas..
- c. Hand washing facilities.
- d. Segregation of pedestrians and vehicles.

4.2 Violence and aggression

There are certain jobs where you are more likely to be at risk of violence and aggression. These jobs usually involve dealing with the public, like security staff, healthcare workers, teachers and police officers.

You are also at a higher risk if you are:
- working alone;
- handling valuable items, including cash;
- working with people who are under stress;
- acting in a position of authority;
- working at night; and
- working in areas with high crime rates.

Some common ways to reduce the risk include:
- reducing cash handling by encouraging the use of credit and debit card transactions;
- reducing customer frustration (introducing ways to reduce waiting times or quickly dealing with customer complaints);
- refusing access to potentially violent customers and clients (perhaps employing security staff who are trained to deal with aggressive and violent people);
- securing doors with entry locks to stop unauthorised access;
- installing surveillance and alarm systems (such as CCTV, panic alarm buttons, observation panels in interview room doors so others can see what's happening and can quickly help);
- improving lighting (so people can't hide in the shadows);
- physically separating staff and customers or clients by high or wide counters, or security screens;
- working/visiting in pairs (particularly if there is a known history of violence or if it is likely);
- training staff (to recognise and defuse signs of violence); and
- issuing personal alarms and mobile phones to use in emergencies.

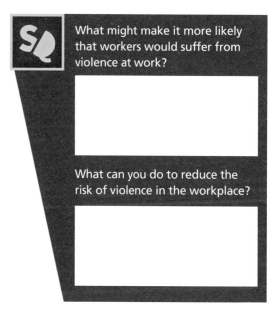

What might make it more likely that workers would suffer from violence at work?

What can you do to reduce the risk of violence in the workplace?

4.3 Work-related stress

At the very least, excessive stress can lead to poor work performance (loss of productivity, lack of motivation, errors and wastage), or behavioural changes. But, if nothing is done, it can lead to increasingly serious health issues such as:

- physical illnesses (including headaches, rashes, stomach problems, sleeping problems, high blood pressure, heart disease, gastrointestinal disturbances); and/or
- psychological or emotional problems (such as difficulty sleeping, anxiety, depression and excessive alcohol consumption).

Many of the things you can do to address work-related stress are simply good management practice. They aren't expensive. Here are just a few things you can try:

- Make sure workloads are manageable – people shouldn't be routinely working very long hours.
- Make sure workers have the right skills and training to do the work and are clear on what is expected of them.
- Give workers as much freedom to plan and organise their own work as you can.
- Involve workers in decisions that affect them (such as working conditions, changes to the organisation).
- Give the right support – such as providing resources where it is needed, taking bullying or harassment seriously and offering counselling when necessary.

4.4 Hazardous chemicals and substances

What is a 'hazardous substance'?

You probably come across hazardous chemicals every day, such as cleaning substances for surfaces or floors, or dust from engineering processes.

Some hazardous substances come in containers – you'll recognise them by the warning labels. For example, cleaning products sold in supermarkets or road tankers carrying fuel. Some hazardous substances are created by a process (such as fumes when welding or dust when sawing wood).

Hazardous chemicals can harm you in different ways. They can also damage buildings and the environment. Helpfully, it tells you this on container labels; for example, it may say that they are toxic, corrosive, flammable or explosive. Sometimes substances are only considered harmful in a particular form. That's because substances have to be able to get into the body (through the skin or eyes, or by you breathing them in or swallowing them). Generally, our lungs are one of the fastest routes into our bodies and so dusts, vapours, fumes, fibres, mists and gases have an easy time getting there. For example, a solid block of wood wouldn't be considered dangerous for your health but in the form of a high concentration of wood dust it certainly would.

The same hazardous substance can also cause different effects; this depends on how long and how often you are exposed to it and what part of the body it is affecting.

For example, if you accidentally splash strong acid on your skin and quickly wash it off, it might just make your skin a bit sore and itchy for a while. If you leave it on for longer, it might cause serious chemical burns. But if you splash it into your eye, it will be very painful and may damage your vision, even if you wash it out straight away.

Some substances may not seem to harm you much at all. In fact, they may seem to be perfectly safe. But if you keep being exposed to them over months or years, serious, irreversible diseases, like cancer, show up. Asbestos (which was widely used in buildings because of its insulation and fire-resistant properties) is an example of this and many builders and electricians are still suffering the consequences.

Hazardous substance controls

Over the years, people have worked out lots of different ways to protect themselves from hazardous substances. These are mostly about stopping the substances from getting into you in the first place. Because some ways are more effective than others, you usually need to use a range of different methods to get the best protection (and still be able to do your job).

Some common ways are:
- Changing the process so that it doesn't need to use the hazardous chemical at all. This completely removes the hazard, such as cleaning the floor with a non-hazardous substance (but it might introduce other hazards). This isn't often realistic but is still worth thinking about.
- Replacing the hazardous substance with either something that isn't hazardous at all or at least something that's safer. Sometimes, it's just a matter of changing the form or the concentration. For example, instead of using the chemical as a powder (which creates a lot of dust that can be breathed in), you might be able to use it as a solution, slurry or pellets where it doesn't so easily create a dust. This can even be done at the point at which the hazardous substance is generated. For example, many power tools for cutting concrete curb stones in road building have an integrated water mist/spray that immediately knocks down most of the dust as it is created.

- Reducing how much time workers are exposed to a hazardous substance. This will reduce the effects it will have on your health. This is easy to do by having regular breaks away from the chemical. For many chemicals, you can also achieve this by sharing the job between several workers, so that each one takes a turn and spends time away doing another job (though this is not recommended with chemicals that cause long-term problems like cancer, as this can mean that more people could end up with long-term health issues).
- Enclosing it (ie putting it in a 'box' or 'pipe' which partially or fully seals it in), which physically separates you from the hazardous substance. This is quite common when manufacturing or transferring chemicals which might otherwise create lots of dust, fumes or vapours for workers to breathe in. Filling your car with fuel is a simple everyday example of this. You don't pour fuel from an open bucket into your car; it's pumped from a large storage tank into your partially enclosed fuel tank through a hose – this greatly reduces how much vapour you are exposed to.

- Using ventilation. Ventilation reduces the concentration of the hazardous substance by diluting it with air. Standard room ventilation (including opening a window) will help; however, if you are creating large amounts of a hazardous substance in the air in a short period of time, you will need something more. In this situation, you need to turn to local exhaust ventilation (LEV). This may sound grand, but it works just like a powerful vacuum cleaner by capturing the dust/mist/vapour/fumes right where it starts, and carrying it away from workers. Many power tools (such as grinding wheels and circular saws) are fitted with these.
- Storing chemicals safely – when you're not using chemicals, keep them in proper containers and away from things they might react with. Only store what you need – stockpiling vast amounts of extra chemicals is just asking for trouble.
- Keeping things clean by having separate 'dirty' and 'clean' areas. This is especially important for highly toxic substances (like lead compounds). Make sure you have eating/drinking areas away from contaminated areas and that workers wash off any contaminants before eating or drinking.
- Using personal protective equipment (PPE) such as gloves, overalls and respirators (face masks that filter out hazardous substances). These are often used together with some of the methods we've already talked about because they can make things even safer (though they shouldn't be relied upon as they are very dependent on people wearing them properly).
- Training workers about the hazards and what to do about them.
- Monitoring concentrations of hazardous chemicals in the air. For some hazardous chemicals, it's a good idea to measure their average concentrations in the air. You can compare these to legal standards to see if you are controlling their concentrations well enough.
- Monitoring worker health. This can help you detect work-related health problems at an early stage before things get worse. It can be as simple as a regular questionnaire about health problems or a regular full medical check, depending on what you're looking for.

Which of these control measures is an example of an 'elimination' control?

Select ONE answer.

- **a.** Using granulated pottery glazes instead of powders.
- **b.** Job rotation.
- **c.** Using water-based paints rather than solvent-based ones.
- **d.** Providing workers with a respirator face mask.

Why is local exhaust ventilation (LEV) preferable to opening a window (dilution ventilation)?

Select ONE answer.

- **a.** LEV speeds up the process.
- **b.** Dilution ventilation cannot be controlled.
- **c.** LEV takes away the hazardous substance at source.
- **d.** LEV is cheaper to run.

Hazardous chemicals and substances

4.5 Laptops and computers

When you use a laptop or desktop computer, you are repeating the same movements over and over again (typing, using a mouse or trackpad), often whilst sitting for a long time. We tend to sit badly too, either over the keyboard or leaning back in the chair. All this can cause stress and strain on your body.

Health issues from computer use include:
- Work-related upper limb disorders (WRULDs) – a collection of problems with arms and hands caused by repetitive use of the keyboard and mouse for long periods of time. Early symptoms of WRULDs often include tingling sensations, numbness and discomfort, but then progress to more severe and long-term pain.
- Back pain – caused by sitting in a fixed position, perhaps with poor posture, for long periods of time.
- Eye strain – temporary tiredness from staring at the screen for long periods of time.

You can make things much better by making sure a computer workstation is set up properly for each user. It's mainly about good posture, comfort and adjustability (adjustable seats and screens, and even adjustable height desks, are common).

You should make the following checks:
- Adjust angle to seat back so that you have good lower back support.
- Adjust seat height so that:
  - your hands have a comfortable position on the keyboard (wrists should be straight and flat when on the keyboard);
  - forearms are approximately horizontal when your hands are on the keyboard: and
  - your feet are supported on the ground (this prevents excess pressure on the underside of your thighs and backs of knees). If you cannot put your feet on the floor, use a footrest to raise them.

- Leave enough space under your desk to fidget and change position while you work (so don't store loads of things under your desk).
- Adjust screen height and tilt so that your head is in a comfortable position.
- Leave enough space in front of the keyboard to support your hands/wrists during pauses in typing; a wrist-rest can provide further support if you need it.

There are now many different types of computers and devices, not just traditional desktop PCs. Being able to work from anywhere with a laptop or tablet is great but it can encourage poor posture (sat on a sofa, having the device in your lap, or even at the wrong height table in a cafe). This is OK for short periods of time, but not for long periods. When using laptops for long periods of time, it's a good idea to treat them more like a desktop computer – sit properly at a desk and plug in a separate keyboard, mouse and screen so that you can get comfortable.

You could use the following checklist to help you: **www.hse.gov.uk/pubns/ck1.htm**

SQ What are the main injuries you can get from using computers?

What should you do to reduce the chance of injury?

4.6 Substance abuse

Signs

If someone is misusing alcohol or drugs they could cause incidents at work. This can be especially bad if they are using dangerous machinery, driving a vehicle or they need to make safety-critical decisions (like an air traffic controller or emergency responder).

You might be able to spot things in a person's behaviour that could lead you to suspect abuse of alcohol or drugs. These are some common signs:
- sudden mood changes;
- unusual irritability/aggression;
- confusion;
- poor relationships with others;
- theft (to fuel an expensive habit) and dishonesty;
- loss of productivity (due to poor concentration); and
- being off work often for short periods.

Before you do anything, it's important to involve medical professionals (who should know a bit more about this than you) in anything that you decide to do. But you should have a clear policy so everyone knows exactly what will happen if they are suspected of abusing drugs or alcohol at work. Whether or not you decide to treat it as a health issue or a disciplinary issue (or both) often depends on the attitude of the person abusing the drugs/alcohol. You may decide to introduce routine drug testing (for safety-critical roles), refer people to counselling or rehabilitation specialists, or even dismiss those who repeatedly flout rules or refuse help.

What might make you think that somebody could be under the influence of alcohol or drugs while at work?

Select ONE answer.

- **a.** Normal behaviour.
- **b.** An unusual change in behaviour or attitude.
- **c.** Being on time.
- **d.** Increased attendance at work.

What effects might the misuse of substances and alcohol in the workplace have on the employer?

Select ONE answer.

- **a.** Increased productivity.
- **b.** Decreased productivity.
- **c.** Happy workforce.
- **d.** Higher profits.

4.7 Electricity

We all use electricity at work, even if it is just to turn on the printer, heat up our lunch or light the workplace. Electricity is vital for our work but it is also dangerous.

Electrical equipment and installations can develop faults, such as a loose wire that becomes exposed or insulation breaking down. This is sometimes through misuse or neglect or just by using it in a damp, dusty, or corrosive environment that it was not designed for. This can mean that normally safe parts (like the casing or handle) can become live, or live parts that are normally covered up become exposed.

Touching these live parts accidentally (or when you expect everything to be OK) can cause electric shock or burns, which can prove fatal. Even if the electric shock itself isn't fatal, it could cause your body to jolt and, for example, fall off a ladder or fall into the path of a vehicle. Faulty equipment can also start electrical fires from sparks arcing or overheating.

The most common area where you might see problems is in portable electrical equipment (such as power tools and kettles). That's because this equipment is taken into lots of different environments and is handled a lot, so it becomes worn or damaged. People are also more likely to be touching the piece of equipment when things go wrong. It is, however, an area that can be easily dealt with because most issues can be found just by looking at the condition of the equipment.

Here are some tips for safe use of electrical equipment:

- Select and use the right equipment for the job and conditions in the first place – think about what it was designed for. If you are working in a rough environment like a construction site, get tools that are tougher, so they can take the abuse. If working in a potentially flammable area, use specially protected (explosion-proof) equipment.
- Make sure that any electrical protection measures are in place, working, and properly used and maintained. Many of these are designed to cut the power in case of fault (either automatically or manually) – for example, fuses, earthing, isolator switches, circuit breakers and residual current devices (RCDs). Some equipment is also 'double insulated'.
- Use reduced or low voltage systems where the risk is greater, for example on construction sites and in wet environments. These systems reduce the severity of electric shock.
- Inspect and test at recommended intervals. For portable equipment, this will usually be a combination of informal user checks (typically a visual check before using it) and formal inspection and electrical tests by a specially trained person. How often you do formal checks depends on the equipment, how old it is and how or where it is used. The manufacturer might also suggest how often. So, for 'portable' things in offices (like a printer) that aren't moved very often (so aren't likely to be damaged), it might be every few years or more. But for things that are used often and in harsh environments, it might be every few months.

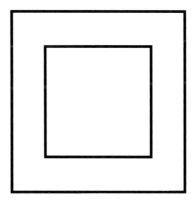

European Symbol used to show double-insulated equipment.

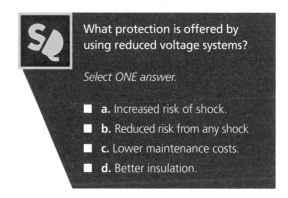

What protection is offered by using reduced voltage systems?

Select ONE answer.

- ■ **a.** Increased risk of shock.
- ■ **b.** Reduced risk from any shock
- ■ **c.** Lower maintenance costs.
- ■ **d.** Better insulation.

4.8 Fire

A fire in your workplace could lead to the building being closed temporarily or being completely destroyed. Some businesses never recover from a fire and workers lose their jobs as a result. Fires can be caused accidentally by all sorts of things but also deliberately (arson). For the most part, deliberate fire starting is the most common cause of fires across workplaces.

EXAMPLE

Individual responsibility

An angry ex-employee of a fast food shop set fire to the premises for the £75 he believed he was due. He set fire to the premises and caused £30,000 worth of damage in the kitchen, though nobody was hurt.

The causes of fire

For a fire to start you need three things:
1. **Heat** – for example from a hot surface, spark or friction.
2. **Fuel** – for example paper, wood or petrol/ gasoline.
3. **Oxygen** – already present in the air but can also be provided by 'oxidising' chemicals or cylinders of oxygen (such as in hospitals or in welding equipment).

These are known as the 'fire triangle': if one of these elements is removed, the fire will go out.

The fire triangle

Classification of fires

There are five categories of fire according to their fuel type. These classification codes vary depending on where you are in the world, but this one is used throughout Europe:

Classification	Characteristics	How it is extinguished			
A	Fires of solid materials such as paper, wood, coal and natural fibres. These fires usually produce embers.	WATER	WET CHEMICAL	FOAM SPRAY	ABC POWDER
B	Fires of flammable liquids or liquefied solids such as petrol, oil, grease, fats and paint.	WATER		FOAM SPRAY	ABC POWDER
C	Fires of gases or liquefied gases such as methane, propane, butane and mains gas.	Gas fires should not be extinguished; this may leave unburned gases as an explosion risk.		Turn off the gas supply.	
D	Fires where the fuel is a metal such as aluminium, sodium, potassium or magnesium.	Special extinguishing media are required depending upon the metal involved.			
F	Fires fuelled by cooking fats, such as deep-fat frying.		WET CHEMICAL		

Heat transmission and fire spread

Fires spread in four ways: convection, conduction, radiation and direct burning.

Convection
Heat rises upwards. This means hot gases from a fire can easily travel upwards in a building (inside or outside) causing the upper floors to get very hot and catch fire.

Conduction
Metals are excellent conductors of heat. For example, if you put the end of a metal spoon in hot water, this part will heat up first and transfer the heat to the rest of the spoon. In a fire, heat is transferred by conduction along metal girders and beams, resulting in the spread of fire throughout buildings.

Radiation
Hot materials give off a lot of radiation. This can transfer heat from one material to another. This means that something can heat up enough to catch fire even if it is not touching the heat source.

Direct Burning
This occurs when a material which is on fire touches another material, causing it to catch fire. A good example of direct burning is of a lit match falling onto a sofa, causing it to ignite.

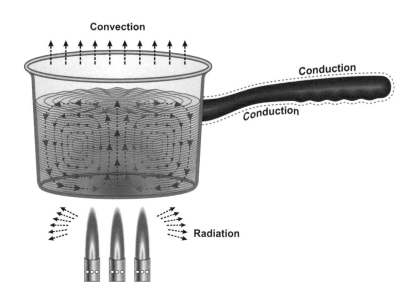

Fire

Fire precautions

Dealing with fire is often broken down into two parts – stopping it from starting in the first place (that's called 'prevention') and limiting how far or how quickly it can spread, making sure people can get away in time (that's called 'protection'). We know from the fire triangle that a fire can't start without a source of oxygen, a source of ignition, and a fuel. So a major part of fire prevention is therefore based on controlling sources of fuel (combustible and flammable materials), potential ignition sources and potential oxygen sources.

Some common approaches to preventing fires are to:
- use and store flammable or combustible materials safely: don't store more than you need to, and keep them away from potential ignition sources or anything that they could react with and start a fire;
- not allow people to smoke in the workplace: smoking should only be allowed in designated areas;
- use explosion-proof equipment in areas likely to contain flammable atmospheres such as paint spraying using solvent-based paints;
- use special procedures to authorise work that carries a higher risk of fire, for example use of 'hot work permits';

An example of a fire extinguisher

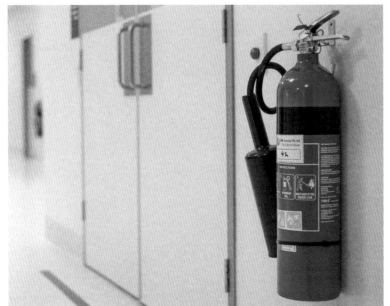

- maintain machinery (eg- use lubrication so it doesn't overheat) and check/inspect electrical equipment (see checks for portable electrical equipment discussed earlier); and
- ensure good housekeeping, for example make sure that litter doesn't build up and any combustibles are tidied away from possible ignition sources. To reduce risk from arsonists, store waste containers far enough away from the site boundary (out of sight and/or reach).

Fire protection is all about getting everyone out quickly and safely and, importantly, giving people as much time as possible while they do.

Common precautions include:
- Fire detectors – these usually work by detecting smoke, heat, flames, or radiation. In some cases they can trigger fire extinguishing equipment directly, like sprinklers as well as raise the alarm.
- Fire alarms – these are usually set off automatically by a detector. There are also manual alarm activation points if someone notices a fire. If you have workers who are hearing impaired, you'll also need to supplement with systems that activate flashing beacons or vibrating pagers (some systems send messages to mobile phones, causing them to vibrate). Alarms may also automatically call the fire department or this may need to be done manually.
- Fire extinguishers can be either fixed (such as sprinklers) or portable. Portable extinguishers are usually sited near exits on fire evacuation routes and close to specific fire risks. You need to use the right type of extinguishing agent for the right kind of fire(s), otherwise it might not put the fire out or it might even make the fire worse. Fire blankets and hose reels are also used.

- An emergency evacuation route – this is a route (or multiple routes) out of the building made up of corridors, fire doors, fire exits, etc. The idea is that these provide a protected route to get out of the building and so should be kept clear. They are designed to stop smoke spreading into them (fire doors and their frames have special seals) and also withstand fire for a minimum amount of time (typically at least 20 minutes). This is why things like fire doors shouldn't be kept wedged open!
- Emergency lighting – normal electricity in a building may fail during a fire, so you need a separate lighting system for emergencies. These are usually used for the emergency route and are commonly combined with emergency exit and directional signs.
- Signs and notices – emergency exit and other signs telling you which way to go to the nearest exit, and to the assembly point. There are also notices telling you 'what to do' if you discover a fire or hear the fire alarm.
- Assembly point – this is usually outside in a safe place where people leaving the building in an emergency will gather until it is safe to go back in.
- Evacuation procedure – a written procedure to tell everyone what to do when they discover fire, hear the alarm, where to assemble, etc. It will also describe special roles such as fire marshals/wardens who help during the emergency (such as taking roll call or checking areas for people left in the building). You should also develop specific procedures for people who might need assistance to evacuate, for example anyone with a mobility impairment, who can't get out as quickly as others.

- Fire training and drills – initial training and then regular practice drills to make sure everyone knows what to do and where to go if there were a real fire.

It's very important to maintain all these systems, for example having annual fire extinguisher checks and fire alarm tests. If you share a building with other businesses, you'll need to make sure that you co-ordinate with each other, or the landlord, so that everyone knows what to do in an emergency.

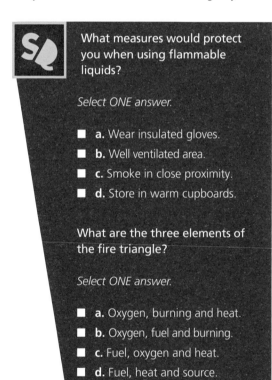

What measures would protect you when using flammable liquids?

Select ONE answer.

- ■ **a.** Wear insulated gloves.
- ■ **b.** Well ventilated area.
- ■ **c.** Smoke in close proximity.
- ■ **d.** Store in warm cupboards.

What are the three elements of the fire triangle?

Select ONE answer.

- ■ **a.** Oxygen, burning and heat.
- ■ **b.** Oxygen, fuel and burning.
- ■ **c.** Fuel, oxygen and heat.
- ■ **d.** Fuel, heat and source.

4.9 Manual handling

Manual handling is an everyday activity for most people. Manual handling happens when you move or support objects at work. For example, lifting and carrying a box of paper to the photocopier or holding an electric drill while you drill a hole in a wall.

Remember that you can injure yourself even when handling light, apparently easy-to-handle objects as well as heavy loads. Manual handling injuries can happen because you:
- use wrong lifting/moving techniques – for example, by holding things too far away from you, twisting or stooping;
- hold loads for too long (or carry a load too far);
- move loads that are too heavy for you;
- use too much force – for example, pulling or pushing too hard or gripping too tightly;
- work at a high pace – repeating the same task many times and not leaving enough time between each to recover properly;
- don't (or can't) grip the object(s) or load safely; *and*
- don't wear the right (or any) PPE.

Common injuries from manual handling include:
- back injuries – slipped or crushed discs;
- muscle injuries – strains/sprains and hernias;
- cuts, scrapes and bruises;
- broken bones (usually fingers); and
- WRULDs – a collection of problems with the hand, arm, wrist and shoulder ranging from pins and needles to serious pain (we looked at these in the section on computers earlier).

Back injury caused by poor posture

Control measures

There are some tried and tested things you can do to make manual handling safer. This depends on looking at your manual handling activity to see if there's a significant risk of injury (picking up a pen or a phone is technically 'manual handling', but injury isn't very likely). If there is a significant risk:

- Where possible, remove the need for manual handling in the first place. Examples include:
  - leaving the load where it is (as it may not need moving at all); or
  - putting loads exactly where they are needed to start with when they are delivered to your site (for example, by using a forklift truck, crane or conveyor to move them from the delivery truck).
- Reduce the risk if it isn't possible to remove the need for manual handling. Most of this can be done by looking at four key areas – the task, individual, load and environment. We will look at each area next.

Task

Think about:

- Using lifting aids such as pulleys, trolleys, sack trucks and pallet trucks but again, make sure these are right for the activity (eg the wrong wheels for a particular type of floor covering or damaged wheels can make trolleys harder to control; if the trolley doesn't have brakes, you'll find it hard to stop going down a slope).
- Altering the order of the activities to minimise the number of operations.
- Reorganising the work layout, reducing carrying distances, avoiding lifting from floor level or the need for twisting or stooping.
- Reducing repetition and introducing variation to provide breaks for muscles to recover.
- Using team lifts or rotating the activities between a number of people.

Individual capability

Make sure that a activity is within a worker's capability (both physically and mentally) – some activities need above-average strength and some workers may be more at risk than others. Think about:

- Those who are especially at risk, for example people with a disability and women who are pregnant – you may need to get special advice from a health professional.
- Giving information and training on manual handling risks and activities – some information (such as weight) are noted on the loads themselves; training is much more effective if it includes real manual handling activities and solutions that are actually encountered in your workplace.

Load

Think about making the loads:

- Lighter/smaller – this means each package carries less risk (not just because of its weight, but also because it has less bulk, so it is easier to see over it). But this may mean that you have more handling operations, with a higher risk from more task repetition.
- Easier to grasp by, for example, providing hand grips or handles. Some things are difficult to hold because they're sharp, hot or slippery, so make sure that the surfaces of the load are clean and smooth, not slippery (or wet, greasy or dusty) and that hot or cold items are held in insulated containers.
- More stable and rigid – consider packing and stacking items differently so that loads don't move when being carried. Think about the centre of gravity of loads. Avoid flexible containers, such as sacks.
- Carry useful information, for example markings that show approximate weight, orientation and where to hold.

Working environment

Think about:

- Workstation design – allow enough space for all movements involved. Workers should be able to position themselves comfortably when lifting, pushing, etc. For example, provide adjustable conveyor belt heights or adjustable chairs for checkout staff.
- Floor conditions – make sure that there are no obstructions, bumps, holes and any materials which may cause people to slip, fall or lose their footing when they are working.
- Changes of level – avoid using steps and ladders wherever possible, when carrying loads in particular (you can trip because it can be difficult to see the steps); ramps are better.
- Working conditions – comfortable temperature (extremes can make you tired more quickly), good lighting.
- PPE –includes gloves, aprons, protective footwear, etc. depending on the risks involved.

What four elements should you consider in a manual handling risk assessment?

Select ONE answer.

- ■ **a.** Task, people, weight, shape.
- ■ **b.** Task, individual, load, shape.
- ■ **c.** Time, individual, load, factors.
- ■ **d.** Task, individual, load, environment.

What is the preferred control measure for manual handling?

Select ONE answer.

- ■ **a.** Change work processes to minimise handling.
- ■ **b.** Remove the need for manual handling by automating the process.
- ■ **c.** Get help with lifts.
- ■ **d.** Lift less frequently.

4.10 Noise and vibration

The effects of exposure to noise

In moderation, noise is harmless, but if it is too loud it can permanently damage hearing. The damage depends on how loud the noise is and for how long you are exposed to it.

You're probably familiar with the effects of loud music at a concert, party or celebration. You can find it difficult to hear properly (things may sound muffled or you might get a ringing sound), but the effects wear off and your hearing returns to normal. However, frequent exposure to excessive noise (like in a manufacturing workshop) can gradually lead to permanent deafness. That not only affects your work, but your social life as well, making it difficult to work out what people are saying, until you might not be able to hear anything at all.

A useful way to tell if you have a noise problem at work is if you:
- have to shout to be understood; and/or
- have difficulty understanding someone who is about two metres away (when you're speaking at your normal conversation volume).

But you may need to measure the levels of noise to properly assess how much of a problem you have and, importantly, work out the best way to deal with the level and type of noise.

Controlling noise

Wherever noise is a problem, you should deal with it in the following order of priority:
- Remove or reduce the noise at the source, for example:
  - electric motors are quieter than diesel engines;
  - plastic buckets are quieter than metal ones when in use; and
  - maintained, lubricated machines are quieter than poorly maintained ones.
- Block the noise coming from the source, for example:
  - enclose the noise source in a soundproof 'box'; and
  - use vibration dampers (eg, rubber pads) to prevent noise from vibration being transmitted to the floor (vibration and noise are very much linked).
- Use noise-absorbing materials in the workroom – thick walls, room dividers, partition walls (with acoustic insulation inside), soft wall coverings can all reduce noise through and within a room. But hard surfaces can reflect noise and make it worse in the room.

Ear muffs

- Provide 'noise havens' for workers which are insulated from noise. This is an alternative to putting the noisy equipment in the box. Instead you put workers in a 'box'. It's useful where you have lots of noisy machines in a room and you might make the control room (for the machines) a noise haven or at least an area where there can be some rest from the extreme noise.
- Install silencers/mufflers on equipment releasing air – the same idea as used on exhaust pipes from car engines.
- Issue workers with hearing protection (ear plugs, ear defenders, etc.) – this should only be used if the level of noise can't be reduced to an acceptable level by other means. This form of noise control is very dependent on workers wearing hearing protection properly. Also, if workers remove their protection for even a short time in a very noisy area, this can effectively mean the damage is done. There is a wide range of hearing protection (including electronic noise-cancelling ones) so you can choose those that fit properly as well as do their job.
- Provide hearing checks – if there is still a significant risk of hearing damage, you should regularly check workers' hearing (it also acts as an extra check that what you have in place is effective). But, obviously, this needs to be done by someone who is trained to do it and knows what to look for (usually an occupational nurse).

The effects of exposure to vibration

Vibration and noise often occur together, especially when working with noisy machines like compressors and pneumatic drills, and in activities producing a lot of vibration such as pile-driving. Vibration is usually divided up into two types:

1. Hand-arm vibration (HAV) – as the name implies, it comes from holding things that are vibrating. Sometimes this is the tool itself when it is used, for example a hammer drill during drilling. Sometimes it's the material that you are holding that vibrates, for example holding material against a grinding wheel. HAV leads to symptoms affecting the hands and arms. Together these are known as Hand–arm vibration syndrome (HAVS), and may be experienced as tingling, numbness and pain in the fingers; blanching/whitening fingers (especially when it's cold); and loss of grip and dexterity.
2. Whole-body vibration (WBV) is usually transmitted through a seat (such as in the cab of a truck driving over rough ground on a construction site) or the floor of a structure – so the floor you are standing on may be vibrating, maybe because there is a large vibrating (and noisy) machine next to you. Regular long-term exposure can give you back pain and poor posture.

Controlling vibration

The approach to dealing with vibration issues is similar to dealing with noise. Here are some for you to think about:

- Treating the vibration at source:
  - use alternative work methods so that you are not coming into contact with vibrating equipment, for example automating the task; and
  - select low-vibration equipment that can perform the task as quickly as possible, reducing exposure time.
- Reducing how much vibration is transmitted:
  - workstation design – provide jigs and holders to reduce the force needed to hold tools, improve posture and reduce the need to grip tools tightly;
  - anti-vibration mountings may also be used; and
  - for WBV through vehicle seats, make sure that the seat suspension is properly adjusted to stop the harsh 'bottoming out' effect when you go over a bump.
- Maintaining equipment – use well-maintained machines to reduce vibration. So make sure that blades are sharp, equipment is lubricated and worn bearings, etc. are replaced.
- Work schedules – control exposure time to avoid prolonged use of vibrating tools such as by job rotation.
- Clothing – provide protective clothing to keep the workers warm and dry, encourage good blood circulation to help prevent HAVS. Using gloves keeps the hands warm but do not rely on 'anti-vibration' gloves to prevent exposure to vibration.
- Health checks – check the health of workers who, despite the controls, are likely to be regularly exposed to high vibration levels. This could be as simple as a regular questionnaire checking for self-reported symptoms such as numbness/tingling in the fingers, to more detailed clinical testing, depending on the extent of the problem.

What are the limitations of ear muffs and ear plugs?

Select ONE answer.

- **a.** They must be chosen by the employer.
- **b.** They rely on a worker fitting/ wearing them correctly.
- **c.** They don't give protection from noise and vibration.
- **d.** They increase noise levels.

How can the risks from exposure to vibration be most effectively reduced?

Select ONE answer.

- **a.** By using higher vibration level tools.
- **b.** By changing ways of working to avoid the use of vibrating tools.
- **c.** By making young workers use vibrating tools.
- **d.** By wearing gloves.

4.11 Work equipment

Key terms

Work equipment
Work equipment is a wide-ranging term that includes machinery, tools (whether hand-held tools or powered tools) or similar, used at work.

Machinery
Machinery is a collection of parts (at least one of which moves) all linked together with all the bits that make it work (like controls, power, etc) to do a particular job. Simple machines are things like pulleys (which might be 'powered' by humans or animals) but more complicated machines are powered tools like drills and lathes.

Hand-held tools
These are tools that are held in the hand and can be entirely manual, for example axes, screwdrivers, hammers, wrenches; or powered, such as portable power tools.

Portable power tools
Hand-held tools that have an external power source (eg electricity, compressed air, liquid fuel, hydraulic and powder-actuated), including anything from an electric screwdriver to a pneumatic drill.

Work equipment presents high risks to the workers who use them. What can happen depends on the type of tool or machine being used.

Most injuries caused by hand-held tools are contact injuries, for example hitting your thumb with a hammer. Other hazards are caused by poor maintenance (eg using a blunt tool) or faults, such as using an axe when its wooden handle is split.

Portable power tools have the same types of hazard as their manually powered alternatives, but we also have present the power source itself and the increased speed and force it brings. Contact injuries may be much more severe so there is a much higher risk when they are used.

When we discuss the hazards of machinery (whether simple or complex, small or large, portable or fixed), experts usually divide them into 'mechanical' and 'non-mechanical'.

Hand-held tool

Mechanical hazards

Mechanical hazards of machinery can result in the following types of injuries:
- crushing (eg when a lift collapses crushing a person underneath it);
- shearing (eg fingers being trapped and sheared off);
- cutting or severing (eg a knife cut);
- entanglement (eg clothing getting caught in the machine);
- drawing-in or trapping (eg being trapped between two moving machine parts);
- friction or abrasion (eg skin making contact with a smooth machine part like a spin dryer or a rough machine part like a belt sander);
- impact (eg being hit by the moving part of a machine);
- stabbing or puncture; and
- high pressure fluid injection (from hydraulic fluid or compressed air lines).

Non-mechanical hazards

Non-mechanical hazards can cause injuries that are not directly related to the machine, but are often caused by malfunctions or machine breakdowns.

They include (with some example harmful effects in brackets afterwards):
- electrical (electric shock);
- thermal (burns, scalds, fire);
- noise (hearing damage);
- vibration (vibration white finger);
- materials and substances such as chemicals or other substances used, worked on or produced (toxic or flammable gases, vapours, dusts);
- radiation (radiation burns);
- neglect of ergonomic principles in machine design (WRULDs);
- slips, trips and falls from neglecting flooring, access, etc; and
- environment in which the machine is operated, such as wind, snow or rain.

A good example of a machine that can cause mechanical and non-mechanical hazards is a paper shredder. The main mechanical hazards associated with the use of a paper shredder are the drawing-in hazard of the rotating blades and the cutting hazard from the blades themselves. The main non-mechanical hazard is electricity, but noise and dust are also present.

Controls

There are lots of things you can do to make using all work equipment safer:

- select, buy and use the right tools for the job;
- train workers – to make sure they use the equipment correctly, know what the hazards are, and know how to make sure the safety devices are in place and used correctly;
- maintain – this can be as simple as sharpening tools but can also be very complicated;
- check and inspect – by taking faulty tools out of the system and either fixing (if possible) or replacing them;
- workers wear the right PPE – such as goggles, gloves and hearing protection. Be mindful of loose clothing that can be dragged into a machine (taking the worker with it);
- make sure that guards and other protective devices are fitted, working properly and used (this might also include emergency stops and brakes to quickly bring things to a halt in an emergency). We will look at guards in a bit more detail below;
- give workers breaks away from the machine (which is a control measure for noise and vibration);
- use reduced voltage systems when using portable power tools in harsher environments (as discussed in the section on electricity);
- make sure non-portable equipment is secured in place (by bolts, clamps, etc.) so that it doesn't move unexpectedly;
- leave enough working space around the equipment so that you aren't tripping over yourself – clear of waste materials, furniture, etc;
- provide good lighting – so you can see what you're doing; and
- clearly identify and mark all work equipment controls, for example operating instructions, maximum and minimum operating speeds, etc.

Guards and protective devices

'Dangerous' parts of machinery are usually dangerous because they are moving quickly (a high-speed rotating blade can do a lot more damage and is moving much more quickly than when it's stopped). So, things like that are usually guarded in some way to prevent or at least limit access. But there can be a conflict – you still need to be able to use the equipment, so you can't totally cover things like a drill bit or a grinding wheel otherwise it won't do the job.

There are many different solutions:

- guarding – fixed guards (that totally or partially enclose, eg, the casing on a portable drill means you don't get caught in the gears inside); adjustable guards (commonly used on things like circular saws);
- interlocks – these cut power or stop the machine quickly when you try to get close to the dangerous part (such as opening an access door or interrupting a 'light curtain');
- push sticks, jigs – designed to hold the workpiece (instead of you holding it) or to push the workpiece into the machine instead of using your hand.

Example of machine guarding

Work equipment

Emergency controls

Powered machines have one or more controls to start and control the speed, pressure or other operating conditions. They also often have extra controls that are used in emergency situations – though this isn't always necessary as it might not stop the machine any quicker or more safely than the normal controls.

Emergency stop controls should only be used in emergencies (and not to routinely stop the machine). Standard emergency stops have a red mushroom-head, push-in button against a yellow background.

Have a look at some of these case studies about guards: **www.hse.gov.uk/agriculture/experience/ machinery.htm**

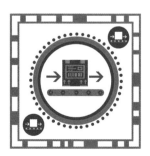

4.12 Work at height

Work at height can be hazardous. But it isn't just a matter of how high up you are (like working 20 metres up the side of a scaffold or skyscraper); it's also about what you can fall onto (or into). It's about falling a distance where you are likely to injure yourself, if you didn't protect yourself. So, this depends on what you're falling onto or into (like a nice soft mattress versus a bed of spikes). That means, it could be as little as a few metres (or even less for the bed of spikes…) before you have to start thinking about taking extra precautions.

Note that work at height can also include working at ground level beside a hole that a worker could fall into, climbing out of an excavation or working beside water.

There are lots of reasons you might need to work at height. For example, fixing a roof or repairing a gutter or even changing a light fitting in a warehouse. Workers falling from height is one of the most common causes of workplace death across the world. Sometimes this is due to not using any (or the right) protection (such as working near the edge of a building roof without a fence or wall to stop you falling). Sometimes, you might be using access equipment (like a scaffold), but it wasn't put up properly, or has become damaged (eg, hit by a truck or weakened by high winds) or overloaded (too much material stored on it). Objects (tools, materials) can also be dropped accidentally (or even thrown), injuring people below.

Work at height

The general approach to work at height is:

- Avoid work at height wherever possible. For example, you might be able to do some of the work on the ground to reduce the time spent working at height (such as prefabrication of items before lifting up into place) or use tools that have an extended reach so you can use them from the ground (like cleaning windows using extending hose pipes). You might even be able to lower things to the ground (some light fittings work like this by design).
- If you can't avoid it, prevent falls from height, for example by using an existing safe place (like a roof with a permanent wall or guard rail already in place around the edge) or using access equipment such as scaffolds, mobile elevating work platforms or work restraint systems.
- If you can't prevent falls, minimise the distance a worker can fall and/or the consequences of the fall, for example by using safety nets, air bags (soft landing) and fall-arrest systems.

It goes without saying that all work at height needs to be done by trained, competent workers (experienced people who know what they are doing).

When trying to prevent falls or minimise distance/consequences of a fall, you should always try to use collective protection (this protects everyone equally) before looking at personal protection (which just protects an individual worker wearing it). 'Collective protection' includes things like guard rails on a scaffold or nets and air bags. 'Personal protection' includes things like work restraints (which restricts how far a worker can travel or lean so that they don't get near enough to the edge to fall) or fall-arrest equipment (which limits how far you fall if you do actually fall).

There's a wide range of equipment used to get to workplaces at height and we've mentioned some of them above. Most types of access equipment will be erected or used by specially trained workers. But ladders and stepladders are two common types of equipment that are very widely used. In fact, you are quite likely to have a ladder or stepladder at home. Ladders don't prevent falls or limit distance/consequences of a fall but they are still great for work that is low risk (eg, light work of a short duration). So here are some quick tips for using them.

AR animation: working at height done badly

Work at height

Safe use of ladders

Leaning ladders

Here are some of the main things to get right when using a leaning ladder. They should:

- be well maintained:
  - not bent or damaged;
  - not contaminated which could make them slippery;
- not be overloaded;
- be long enough for the task;
- be set up on a solid, flat base;
- be secured at the top whenever possible; *and*
- be 75° to the horizontal or at a ratio of 1:4 distance away from the wall to height.

Make sure that:

- only one person is on the ladder at any time;
- you have three points of contact most of the time (normally two feet and a hand);
- you do not overreach – a good rule of thumb is to ensure that your belt buckle (navel) stays within the stiles – and keep both your feet on the same rung or step throughout the task;
- if possible, nothing is carried in your hands while climbing;
- wooden ladders are not painted as this can hide defects;
- ladders are not used near unprotected live electrical lines;
- you don't work off the top three rungs (steps); and
- you try to make sure the ladder extends at least 1 m (three rungs) above where you are working.

Stepladders

When using stepladders many of the same things need to be considered. Clearly the point about lean angle does not apply as this will already be set on a stepladder.

Here are some additional considerations needed for stepladder use:

- do not work off of the top two steps unless they have a safe handhold;
- ensure there is space to fully open them; and
- use any locking devices.

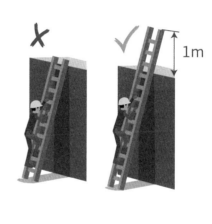

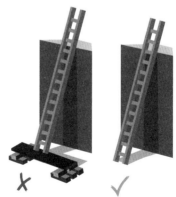

AR animation: working at height done well

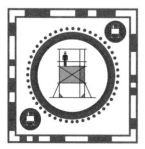

Discuss which of these could be working at height:

- Climbing a stepladder.
- Climbing a staircase.
- Working beside a pier.
- Working at ground level.

4.13 Workplace transport

Workplace transport includes any vehicle used in the workplace, whether it's a delivery van, car, forklift truck or even mobile work equipment. A lot of incidents involve workplace transport; some of the most common (and examples of how they can happen) are:

- people being hit by a moving vehicle (eg a delivery truck hitting a pedestrian while it is reversing, perhaps made worse by bad weather and too much congestion on the site);
- workers falling from a vehicle (eg while standing near the edge on the loading bed of a flat-bed truck or even standing on top of the load itself);
- materials and equipment falling from a vehicle (eg load not being secured properly); and
- vehicles overturning (eg a forklift truck driving too fast around a sharp bend or across a steep slope with the load/forks raised).

Control measures

Many of the control measures for workplace transport are similar to the approach used on public roads. This includes:

- good design of site traffic routes;
- providing suitable, maintained vehicles with driver protection systems;
- having trained and competent drivers;
- producing procedures for high-risk operations, such as loading and unloading; and
- having site rules.

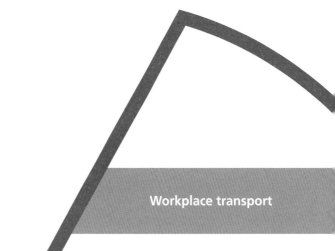

Workplace transport

Design of site traffic routes

Make sure that vehicle routes:

- Are designed to minimise reversing, for example by having one-way systems and turning circles. You can also install traffic lights to control busy areas or entrances, such as around doors.
- Are clearly marked, for example white lines indicate lanes, junctions and loading bays; diagonal yellow lines on a black background clearly mark hazards and identify obstructions such as low bridges, speed bumps, etc.
- Separate pedestrians and vehicles, as far as possible – ideally use physical barriers, but if not possible, mark the road surface to show the separate vehicle and pedestrian routes (especially near exits from buildings which lead directly to a road). Pedestrian crossings (eg, zebra crossings), bridges or subways will also keep vehicles separate from pedestrians. Always provide separate pedestrian access at doorways, so people don't have to walk through a vehicle access area and risk colliding with a vehicle coming out as they enter.
- Are wide enough for vehicles, including passing, turning and loading/unloading areas.
- Are strong enough for the vehicles using them (eg, heavy goods vehicles).
- Have clear signs – directional (such as a one-way system), warning and prohibition signs indicate alerts to drivers. Information and warning signs should conform to international standards (drivers will already be familiar with these, so it makes sense to use them).
- Provide enough parking for all vehicles on site, including cars. Parking in unauthorised areas may cause safety issues, such as by restricting visibility or blocking access.

- Have good visibility – make sure there's enough lighting (especially at junctions, pedestrian areas and where roads are close to building exits) and use mirrors to improve visibility at blind bends, blind exits and sharp corners.
- Provide sufficient space to enter and exit the site (including alternative routes for high-sided vehicles where there is an obstruction such as a low bridge).
- Have speed limits – this may be by speed limit signs but also by physical features such as speed humps, chicanes and rumble strips.
- Are not excessively sloped (because of the risk of vehicles overturning); surfaces should be generally even, well-drained and maintained.

Vehicles

Vehicles should be:

- Fitted with driver protection – such as seat belts, secure doors, and protective cages and cabins with shatterproof glass. Special measures to prevent vehicles overturning or rolling away may also be provided (rollover protection systems). These include safety stops to prevent vehicle movements on slopes, and extendable legs to provide stability when operating. Safe access to and from vehicles is also important (like handrails and footholds).
- Fitted with warning/flashing lights and alarms to indicate movement, particularly on reversing.
- Maintained – paying special attention to braking systems, steering mechanisms and tyres (good tread depth and correct pressure).

Driver competence and selection

Many of the control strategies we have considered rely on the driver operating the vehicle correctly and obeying rules.

Select new drivers carefully. Consider their age, reliability/responsibility, physical fitness and intelligence.

Driver training should cover:

- **General basic training** – the basic skills and knowledge needed to safely operate the type of vehicle and any attachments that the driver will be required to use. For example, a forklift truck driver might attend a training course run at the manufacturer's premises, so that they know how to operate that type of machine correctly.
- **Specific job training** – covering knowledge of the workplace, any special requirements of the work and the use of any special equipment.

Refresher training is also useful for all drivers, but may be specifically required when, for example, drivers must operate different types of vehicles or if there is a change of work.

Special procedures for vehicle reversing, loading and unloading

Special procedures include:

- Reversing – use people to direct vehicle movements if vehicles need to reverse or operate in restricted spaces such as loading bays. Make sure that standard hand and directional signals are used for this procedure.
- Loading and unloading:
  - use a forklift truck for off-loading (to avoid workers needing to climb onto the load area) or, if this isn't possible, provide a safe means of access; and
  - properly secure and distribute the load evenly to prevent movement while in transit.

Site rules

It is important that all drivers and pedestrians using a site are aware of all the rules that apply. Visiting drivers may need to be given both general and specific instructions about the management systems operating on site.

Examples include:

- stick to speed limits;
- keep keys in a secure place when a vehicle is not in use;
- switch off the engine and apply the brakes at the end of the work period. Disconnect the battery on battery-operated vehicles;
- park all vehicles in a safe place and do not obstruct emergency exits, other vehicle routes, fire-fighting equipment or electricity control panels;
- don't leave vehicles unattended on a gradient; and
- when pedestrians and vehicles can't be separated, consider asking pedestrians to wear high-visibility jackets.

Identify things to be considered when developing a suitable traffic management plan for a large DIY/hardware store. The store is open to the public and uses vehicles such as forklift trucks inside the building and yard.

It is important that employers look at all circumstances in which vehicles could cause harm in the workplace, and provide appropriate measures to keep workers safe where vehicle cannot be avoided.

Keeping an eye on how things are going

Element 5

Making the right choices

A worker cuts his finger while lifting a box and walks straight across a vehicle production area to access the first-aid box. He gets hit and seriously injured by a moving vehicle.

Organisations and individuals can often get distracted and lose sight of what are the most significant risks within the work environment.

Example: Making the right choices

5.1 Ignoring the trivial and focusing on the important

It is very tempting to try to do too much. But you can't do everything at once. The challenge is to focus on the most important things first and not the seemingly urgent (but less important or even unimportant). There is no point trying to launch a poster campaign on improving worker attitudes to safety, when you can't even get the basics right. If you are focusing on a campaign for a low-risk issue and ignoring the shocking state of your chemical factory (that could blow up anytime), you've got it wrong. That's just a matter of priorities and applying proportionate resources.

The same thing applies when checking health and safety precautions and processes in the workplace – like when doing an inspection (which usually looks at physical things like the workplace or equipment) or audit (which tends to look at management processes). It's very easy to come up with a huge list of things that are wrong (or could be improved) and get bogged down in activities to fix them. But, you will often find that fixing only a few of the most important issues will make by far the biggest difference to worker health and safety. The remaining issues may make only a very small (if any) difference to worker health and safety if you fix them or not. That doesn't mean to say you just leave them but it does mean you need to look at the potential impact of leaving or fixing things and not just see everything as equally important.

Ignoring the trivial
and focusing on the
important

In this element, we'll look at two important methods of checking – inspections and audits – and what we can learn from them. We introduced these tools in Element 1. Here, we look more about how to use them effectively.

Methods could include inspections, using checklists, safety sampling, auditing a process or a simple walk round, looking at escape routes and housekeeping.

5.2 Inspecting the workplace

You can use a checklist to help you inspect a workplace or a piece of equipment. A checklist could be created by looking at where previous incidents have occurred, or by spotting a trend in reported incidents may lead you to focus a checklist in that key area. It's also good to create a plan for carrying out inspections so you don't just do them once. This can

also indicate which part of the workplace or piece of equipment you'll be inspecting.

Here is an example of an inspection checklist (this is based on the 'Sample Checklist for Offices' from the Canadian Centre for Occupational Health and Safety):

Inspectors:	Date:		
	(O) Satisfactory (X) Requires action		
	Location	Condition	Comments
Floors			
Is there loose material, debris, worn carpeting?			
Are the floors slippery, oily or wet?			
Stairways and Aisles			
Are they clear and unblocked?			
Are stairways well lit?			
Are handrails, handholds in place?			
Are the aisles marked and visible?			
Equipment			
Are guards, screens and sound-dampening devices in place and effective?			
Is the furniture in good repair and safe to use? Look for:			
- chairs that are in poor repair			
- sharp edges on desks and cabinets			

There are advantages and disadvantages to using a checklist when doing an inspection.

Advantages:
- Consistency between inspections, looking at the same things every time;
- Knowing what you should be looking at before you begin the inspection; and
- Not 'forgetting' any parts or miss anything out.

Disadvantages:
- Something could be missed because it wasn't on the list; and
- Could become a 'box-ticking' task if you get too used to what is on the list.

Talking to people

Interviewing or talking with workers (perhaps during an inspection) is a very effective way of checking:
- what workers know about the processes they are following; and
- how they feel about them, for example if they feel that they create a safe working environment.

Workers who are involved in the development and improvement of processes are more likely to understand and follow them.

Inspecting the workplace

5.3 Auditing – what, why and how

Audits are used to find out if there are appropriate management systems in place to manage health and safety. They also check if suitable risk control systems and workplace precautions are being used and are working. For example, you could choose to audit the statutory inspections in the workplace to check whether forklifts, lifting equipment and accessories are being inspected in accordance with legal requirements, and that all documentation is kept.

Health and safety audits are used to determine compliance with an agreed standard. This could be an external standard that the organisation has agreed to meet; it could be the standard set by their internal management system; or a combination of both. Essentially, an audit is checking that the system is doing in practice what it has said it will do on paper.

An audit can be done by a trained member of staff (internal), or by someone from outside the company (external). An external auditor is often a representative of an awarding agency, who knows a lot about the standard which needs to be met.

An external audit might be taken more seriously by senior managers but they can be very expensive and time-consuming. An internal audit can take less time to complete, as it may only look at a small part of the system. However, an internal auditor might be too familiar with the workplace and therefore miss things or assume knowledge without being able to see any evidence. It is likely that both internal and external audits will be carried out on a health and safety management system at different times.

Audits are part of the continual improvement process. They are a proactive way to improve performance, not a negative review looking for failures. Some audits are required by certification bodies; others are implemented as a best-practice approach to improve overall safety. Whatever the reason for an audit, the intent is the same: to identify areas for improvement that can then be used in new health and safety plans.

For an audit to be thorough and effective, you should consider using a variety of methods to find evidence of how the system is working:

- **Looking at documentation** – such as policies, risk assessments and maintenance records.
- **Interviews** – speaking to workers about the processes they follow.
- **Observations** – watching the work activities to see whether the processes are being used and whether they effectively keep people safe.

The three most important things to consider when planning an audit are:

1. **Scope** – what are you looking at. This can depend on what time you've got and how big the organization is (whole systems or highly focus areas/topics).
2. **Objectives** – what are you looking for.
3. **Criteria** – what are you comparing it against.

A common approach is 'risk-based auditing'. Higher risk processes (the bits you are most concerned about) are audited more frequently.

What are the three main methods that might be used during an audit?

Select ONE answer.

- ■ **a.** Looking at documents, checking maintenance records, carrying out an inspection.
- ■ **b.** Looking at documents, interviewing workers, noting observations.
- ■ **c.** Interviewing workers, taking statements and reviewing feedback.
- ■ **d.** Reviewing feedback, addressing complaints and checking maintenance records.

Auditing – what, why and how

5.4 Learning from good and bad practice

It is important that we use the information gathered from inspections and audits to make continual improvements to work activities. Audits will show us which processes are working and which ones are not. We can use this information to learn how and why things are going wrong to prevent a recurrence, and how and why things are going well.

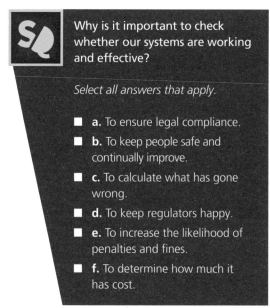

Why is it important to check whether our systems are working and effective?

Select all answers that apply.

- ■ **a.** To ensure legal compliance.
- ■ **b.** To keep people safe and continually improve.
- ■ **c.** To calculate what has gone wrong.
- ■ **d.** To keep regulators happy.
- ■ **e.** To increase the likelihood of penalties and fines.
- ■ **f.** To determine how much it has cost.

Think about what processes you could improve in your organisation.

If your organisation has a formal management system, you could use your organisation's auditing checklist to determine what should be looked at in more detail. If it doesn't, a good place to start is by using the OSHA 'SHP Audit Tool', available online.

Whether you are learning from your own organisation or from the great ideas or mistakes of others, it is important to take these forms of review seriously, so that you can continually improve health and safety within your workplace.

You will find that the process of checking how you are doing will help improve safety overall.

Glossary of key terms

Accident
An incident which result in damage, death, injury or ill health.

Audit
An audit is a check of processes, procedures and systems, so will compare what should happen (according to a process/procedure) with what actually happens. This is a common active monitoring method.

Control measure
Anything that you do or put in place that eliminates or reduces the risk of harm or damage.

Hand-held tools
These are tools that are held in the hand and can be entirely manual, for example axes, screwdrivers, hammers, wrenches; or powered, such as portable power tools.

Hazard
Anything that has the potential to cause harm or damage – this could be an object, an activity, or even a situation, or a combination of these.

Hazards are often described either by referring to the type of harm or effect they lead to (eg, mechanical machinery hazards such as crushing, entanglement or even slips, trips or falls) or instead by the hazard origin or source (eg, electrical or noise).

Immediate cause
The obvious cause(s). For example, the exposed blade on a machine that leads directly to a cut hand.

Incident
An undesired event that has caused or could have caused damage, death, injury or ill health. There are two main types of incident – accident and near miss.

Inspection
A physical check of a workplace (such as an office or warehouse) or machine, often using a checklist of hazards/issues to look for. This is a common active monitoring method.

Machinery
Machinery is a collection of parts (at least one of which moves) all linked together with all the bits that make it work (like controls, power, etc) to do a particular job. Simple machines are things like pulleys (which might be 'powered' by humans or animals), but more complicated machines are powered tools like drills and lathes.

Manual handling
Transporting a load using muscular strength and body weight. It may involve lifting, pushing, pulling, carrying, lowering and supporting of the load. The load can be anything, for example a box, a hand drill or an animal.

Mechanical hazards
Hazards arising from the direct interaction of people with the machine itself.

Near miss
An incident that could have resulted in damage, death, injury or ill health but did not, in fact, do so.

Non-mechanical hazards
Associated with the use of machines, often the environment within which the machines are located, the materials used and other aspects of the machine's operation.